TABLE OF CONTENTS

STEP ① USING THIS BOOK

This Book Was Designed for You —

a talented, yet busy teacher. We know that you want to provide students with the most interesting and comprehensive units of study possible. We also know how much time it takes to fully prepare to teach a topic. That's why we developed the **Ten Easy Steps** series. From planning to implementation — it's all here.

STEP ① Using This Book
This section contains a little background information on how to use this book and a peek into what you'll be teaching during the ten lessons.

STEP ② Gather Great Resources
In this section, you'll find a list of books to use when teaching yourself and others about magnets and electricity, a list of web sites that help explain the topics you'll be teaching, and a list of field trip and guest speaker ideas. There's even a letter for parents to help you find a great speaker!

STEP ③ Speak the Lingo
This is where you'll find all the vocabulary words and definitions specific to the topics covered in this book as well as worksheets and pocket chart ideas designed to reinforce the vocabulary.

STEP ④ Set the Scene
It's important to set the tone for the unit of study. This means transforming your classroom environment to reflect the concepts being taught. In this section, you'll find great ideas for interactive learning areas and classroom decoration.

STEP ⑤ Plan a Project
In this section, you'll find plans for an ongoing project students will be working on throughout the unit of study. It's a great way to apply what they're learning each day.

STEP ⑥ Teach Ten Terrific Lessons
Ten complete lessons can be found within this section. Each lesson includes essential concept information, experiments, hands-on activities to reinforce the concepts, journal prompts, homework ideas, and teaching notes on each experiment.

STEP ⑦ Cross the Curriculum
Take one great concept, teach it in multiple curriculum areas, and you're sure to reinforce learning. In this section you'll find ways to extend the learning across all areas of the curriculum, including social studies, reading, writing, math, and art.

STEP ⑧ Tie in Technology
In this section, we provide you with ideas and project planning pages for a multimedia presentation and web site creation.

STEP ⑨ Assess Learning
This section provides a variety of assessment options. But don't wait until the end of the unit to assess your students. This book is filled with journal and homework ideas to assess from the start.

STEP ⑩ Celebrate!
Once you've completed a unit study as compelling as this, you'll want to celebrate. In this section, we've provided an idea for a great end-of-unit celebration.

A Note About the Internet
The Internet is a constantly changing environment. The sites listed as additional references were current at the time this book went to press.

Electricity is so much more than flipping a switch to turn on a blender. Electricity is a science that involves a phenomena at the atomic level, where electrons are busily moving about. Controlling electrons' movement is what produces electricity. Magnetism is also an amazing phenomenon. Magnets are at work in every electrical motor we use and in many objects around us.

In this unit of study, students will learn the basics of electricity, including the structure of the atom with its electrons, how to make a battery, the three types of circuits, the difference between conductors and insulators, and sources of electricity. Students will also learn about the characteristics of magnets by making a lemon magnet, magnetizing iron, and finding out more about our biggest magnet, Earth.

Each of the following lessons in **Step 6** features a quick, informative mini-lesson, easy-to-accomplish experiments and activities, a journal prompt, and a homework idea.

LESSONS

1. Electrons: The ELECTR in Electricity
Objective: To learn about the atom and how electrons move around it.

2. Current Electricity
Objective: To learn how to make a homemade battery and principles of how it works.

3. Conductors vs. Insulators
Objective: To learn which substances conduct electricity and which do not.

4. Making Connections: The Three Basic Circuits
Objective: To build and illustrate simple, series, and parallel circuits.

5. Electrical Energy Sources
Objective: To analyze and critique the various resources that are used to produce electricity.

6. What Makes a Magnet a Magnet?
Objective: To learn about the characteristics of magnets by investigating how they work.

7. Magnetize/Demagnetize
Objective: To explore ways to magnetize and demagnetize iron objects.

8. The Earth as a Magnet
Objective: To learn how a compass reacts to Earth's magnetism and build a homemade compass to test.

9. Electromagnets
Objective: To learn about electromagnets and to build one.

10. Electric Motors
Objective: To learn how an electric motor works by building one.

OTHER TOOLS

In addition to the great lessons and experiments, this book also contains all the other tools to help you make your unit on magnets and electricity more complete, including:

- A list of books and web sites for you and your students. **(Step 2)**
- A vocabulary list and definitions along with vocabulary worksheets, puzzles, and pocket chart activities. The back of the book contains a pocket chart card for each vocabulary word. You can use the pocket on the inside back cover to store the cards once they're torn out from the book. **(Step 3)**
- Learning center ideas filled with information to help you set up an electricity and magnetism discovery center. **(Step 4)**
- An ongoing project in which students host an *Electricity and Magnetism Invention Quest.* **(Step 5)**
- Cross-curricular learning ideas to carry the study of electricity and magnetism into other areas of your curriculum. This includes graphing energy consumption of households; analyzing electricity costs in other states/other countries and their dependence on various electricity sources; composing a rap or poem; researching the first magnets; and looking at the impact that magnets and electricity have on other countries and cultures. **(Step 7)**
- Integration into technology with an electricity and magnets web page project and a multimedia presentation about conserving energy. **(Step 8)**
- Assessment tools including rubrics, journals, and tests. You'll find plenty of tools and ideas for alternative or traditional assessment of student learning. **(Step 9)**
- A celebratory end-of-the-unit event that allows students to "show what they know" while reinforcing the content covered. **(Step 10)**

STEP ②
GATHER GREAT RESOURCES

Great Resources for You

It's impossible to be an expert on every subject you teach, yet that's exactly how your students see you. Before you begin teaching this electricity and magnets unit, spend a few nights reviewing the following web sites and books and you'll be up to speed in no time!

Web Sites

Electricity and Magnetism
http://www.galaxy.net/~k12/electric/index.shtml

This site lists and gives directions for electricity and magnetism experiments.

Internet Guide to Electronics
http://webhome.idirect.com/~jadams/electronics/

This site is a good place for beginners to learn about the basics of electronics.

Articles on Electricity
http://www.eskimo.com/~billb/ele-edu.html

This site has all the answers to kids' and adults' questions about electricity.

Spectroscope of an Atom
http://www.miamisci.org/af/sln/phantom/spectroscope.html

This site gives a good visual and written explanation of the atom.

Magnet Man: Cool Experiments with Magnets
http://search.britannica.com/frm_redir.jsp?query=magnet&redir=http://www.execpc.com/~rhoadley/magindex.htm

This is a good overall site for learning about magnets.

Learning Resources
http://www.learningresources.com

Seek out this site for a list of **10 Steps**-recommended web sites or great products for your classroom. You'll want to head to *Activities & Resources* for the list.

Books That Help Prepare

Asimov, Isaac. *Understanding Physics: Light, Magnetism and Electricity.*
New York: New American Library, 1966.

This is one of the best books for people who may not have had any physics in high school. The concepts are explained in a clear manner with wonderful illustrations to guide through them.

Chapman, Philip and David Crawley. *Usborne Young Scientist: Electricity.*
London: Usborne Publishing Ltd, 1991.

This book explains what electricity is, how it works, and how we use it. It also discusses how electricity is made and transmitted around the country and includes safe and simple experiments.

Cooper, Alan. *Electricity (Visual Science Series).*
Morristown, New Jersey: Silver Burdett Company, 1985.

Using a strong visual approach, detailed artwork, diagrams, and colorful photographs, this book helps make the concepts of electricity easy to understand.

Levenson, Elaine. *Teaching Children About Science.*
New York: Prentice Hall Press, 1985.

This book is written for parents and teachers who would like more background materials for teaching science content to children. It helps to answer the questions "why" and "how" and assists parents and teachers in enhancing critical thinking skills in children.

Great Resources for Your Students

Surrounding your students with great resources is a sure way to stimulate learning. The first step is to encourage your students to take a look at a few of the great web sites and books listed on this page and page 7. The field trip ideas in this section will also get your students in gear for a great study of magnets and electricity. You'll have a captive audience before you even begin teaching!

Web Sites

The Energy Story: What is Electricity
http://www.energy.ca.gov/education/story/story-html/chapter02.html

This web site is connected to others all with great information about electricity.

The Atoms Family
http://www.miamisci.org/af/sln/index.html

Monsters guide you through a lot of scientific principles.

Snacks about Electricity
http://www.exploratorium.edu/snacks/iconelectricity.html

This is a good site that explains a lot of general principles about electricity.

Articles on Electricity
http://www.eskimo.com/~billb/ele-edu.html

This site provides links to other sites to help explain the phenomenon of electricity.

Science: How the World Works
http://www.brainpop.com/science/forces/magnetism/index.wem

This site offers movie clips of many different science concepts.

Teach Learn Communicate
http://www.alfy.com/teachers/teach/thematic_units/Electricity_Magnetism/EM_1.asp

This site provides good short movies on electricity and magnetism.

Marshall Brain's How Stuff Works
http://www.howstuffworks.com/battery.htm

This site explains in detail how a battery works.

Magnet Man: Cool Experiments with Magnets
http://search.britannica.com/frm_redir.jsp?query=magnet&redir=http://www.execpc.com/~rhoadley/magindex.htm

This is a good overall site for learning about magnets.

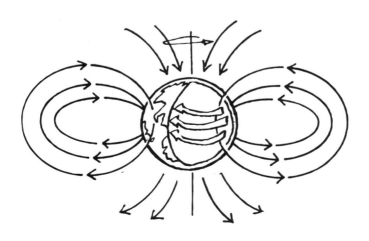

Great Resources for Your Students

Books

Amery, Heather. *Knowhow Books of Batteries & Magnets.*
London, England: Usborne Publishing, Ltd., 1989.
This book provides a collection of simple to make gadgets for exploring magnetism and electricity. It also includes "you will need" shopping lists for each activity, easy-to-follow instructions, and many interesting experiments.

Graf, Rudolf F. *Safe and Simple Electrical Experiments.*
Mineola: Dover Publications, Inc., 1964.
With 101 entertaining experiments and projects, this book offers a fast and reliable way of learning the basic principles of electricity. Experiments use safe and inexpensive materials, and they're accompanied by detailed step-by-step instructions and illustrations plus a brief discussion of the results.

Gutnik, Martin J. *Simple Electrical Devices.*
New York: Franklin Watts, 1986.
This book takes a close look at circuits, batteries, lights, motors, generators, the telephone, tubes, and transistors. It also includes easy projects and experiments.

Notkin, Jerome J. and Sidney Gulkin. *How and Why Wonder Book of Electricity.*
Los Angeles: Price/ Stern/Sloan Publishers, Inc., 1987.
This book, in story form, explains how one family answers their children's questions about electricity. Each experiment is accompanied by excellent illustrations, instructions, and "how to and why" information.

Walker, Jearl. *The Flying Circus of Physics with Answers.* *New York: John Wiley & Sons, 1977.*
This original, offbeat book is a collection of problems and questions about physics in the real, everyday world. The questions focus on relevant and fun phenomena like the sound of thunder, rainbows, sand dunes, and soap bubbles. It also includes a separate section on electricity.

Vecchione, Glen. *Magnet Science.*
New York: Sterling Publishing, Co., 1996.
This is a very good book to use when learning about magnets.

Guest Speaker Ideas

1. A local electric company engineer or electrician.
2. A telephone company technician (the person who works on the telephone poles).
3. Someone who has been struck by lightning or experienced the effects of lightning.
4. Someone who works at a hydroelectric dam.
5. Someone who uses and understands solar power.
6. Someone who builds or repairs electric motors.
7. An electrical engineer who knows about magnets and electricity.
8. A meteorologist to explain the science behind lightning.

Field Trip Ideas

1. Visit the local electric company.
2. Visit a hydroelectric dam or a coal, oil, or natural gas factory.
3. Visit an appliance store for an "electricity scavenger hunt," discussing how electricity is related to the items sold in the store.

GATHER GREAT RESOURCES

Letter to Parents

Dear Parents:

Over the next few weeks our class will be studying electricity and magnetism. Our topics of interest will include:

1. **Electrons**
2. **Current Electricity**
3. **Conductors vs. Insulators**
4. **Making Connections: The Three Basic Circuits**
5. **Electrical Energy Sources**
6. **What Makes a Magnet a Magnet?**
7. **Magnetize/Demagnetize**
8. **The Earth as a Big Magnet**
9. **Electromagnets**
10. **Electric Motors**

If you have personal stories or insights to share on any of the above listed topics, we would love to have you come in and talk to the class. We would also appreciate any materials (books, videos, and posters) that you'd be willing to share for the next few weeks.

Reinforcing learning at home will help your child retain the information learned in school. Try to find time to discuss the topics, ask questions, and stay involved with homework and projects. If possible, explore the following web sites with your child.

Electricity and Magnetism
http://www.galaxy.net/~k12/electric/index.shtml

Internet Guide to Electronics
http://webhome.idirect.com/~jadams/electronics/

Thank you for all your help and support.

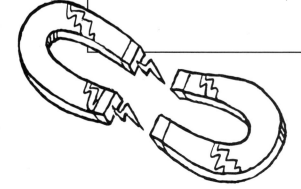

STEP 3
SPEAK THE LINGO

Electricity and Magnets Vocabulary

Understanding the meaning of key words before delving into a topic will help students grasp the concepts later on. The pages in **Step 3** provide the practice to help students retain the words and their definitions. The worksheets are based on the following list of vocabulary words, which are from the lessons in **Step 6.** Each word is printed on the pocket chart cards located at the end of this book.

Lesson 1

atom
smallest unit of all matter

electricity
collection of electrons (static electricity) or movement of electrons (current electricity)

electron
negatively charged particle that orbits the nucleus of an atom

neutron
part of an atom's nucleus that has no charge

nucleus
central part of an atom

Periodic Table of Elements
lists all the elements or substances humans have discovered so far

proton
positively charged part of an atom's nucleus

static electricity
caused by non-moving, electrical charges that have built up in one place

strong force
holds positively charged protons along with neutrons in the nucleus of an atom

---

Lesson 2

alternating current (A.C.)
current that flows back and forth in a circuit (such as power from an electric company)

ampere
unit of measure of current through a circuit

---

battery
acts like a "pump" that forces electrons through a conductor

conductor
material that easily allows an electrical charge to move through it

current
flow of electrons or the amount of electricity flowing through a conductor at a given time

direct current (D.C.)
current that flows in one direction in the circuit (e.g., power from batteries)

volt
unit of electric potential

Lesson 3

circuit
path formed by the conductors for electrons and contains an energy source, conductor, receiver, and resistor

filament
causes the light in a light bulb

insulator
material that does not easily allow charges to move through it

resistor
device that slows the flow of electrons

resistance
measure of how difficult it is to move electrons through a conductor

---

short circuit
created when electrons pass from the negative to the positive pole of a battery without traveling through a resistor

---

Lesson 4

parallel circuit
created when there are two or more separate paths or branches for electrons to follow and the voltage is the same in each branch

schematic drawing
drawing that uses symbols to represent objects

series circuit
ype of circuit in which there is only one path for electrons to flow and there may be two or more resistors (if the circuit is broken, the current in all parts is stopped)

simple circuit
circuit in which there is only one resistor

Lesson 5

No new vocabulary

Lesson 6

magnet
any object that has a magnetic field and is able to exert force on other magnets

magnetism
physical property of matter in which there is an attraction due to unlike poles

Magnetic Resonance Imaging (MRI)
technique that uses electromagnetism to activate atoms in the body and produce computerized images of tissues

natural magnet
magnet made of lodestone or magnetite, both of which are naturally magnetic

permanent magnet
magnetic metal that has been magnetized and will stay magnetized for a long time

temporary magnet
material that has been magnetized by rubbing a permanent magnet on it and maintains a magnetic field for a short time

Lesson 7

demagnetize
to scramble the atoms in a temporary magnet

magnetic field
area of magnetic lines of force

magnetize
to create a magnetic field in a piece of iron by lining up the atoms using a permanent magnet

Neodymium-Iron-Bar (NIB) magnet
very strong super magnet

Lesson 8

compass
suspended magnet with one end pointing toward Earth's magnetic North Pole

magnetic North
point on a compass at which the needle points North, in line with Earth's magnetic field, and is slightly shifted from true North due to Earth's tilt

true North
North Pole

Lesson 9

electromagnet
piece of iron wrapped with copper wire that has a charge going through it.

Joseph Henry
built the first electromagnet

Lesson 10

deflecting force
force a magnet exerts on a current carrying wire, used to drive most electric motors

Teaching Notes on Pocket Chart Vocabulary

Using your pocket chart cards and a pocket chart, try a few of the activities listed below to strengthen vocabulary skills.

Begin Each Lesson

Begin each lesson by showing the new vocabulary words that apply for that lesson. At the end of each lesson, review the words with your students together.

Quiz Show

Play "Electricity and Magnetism Quiz Show." Divide the class into teams, pull one vocabulary card, and give its definition without showing the face of the card (example: a path formed by the conductors for electrons to move). The first team to "buzz in" with the correct word (circuit) receives a point. Continue until all the cards have been revealed.

Definition, Please

Play "What's the Definition, Please?" Place all the cards facedown in the pocket chart. Divide the students into four teams. Teams take turns sending a player up to the chart to retrieve a card and take back to their group. The group then has 30 seconds to come up with a definition for the word to receive a point. If the group cannot come up with the definition, the other teams have the opportunity to answer. The first team to "buzz in" with the correct definition gets the point, and regular play resumes with the next team going up to draw a card. Continue until all the terms have been defined.

Electricity vs. Magnetism

Play a round of "Electricity vs. Magnetism" with your students. Select a card and have students write down whether it falls under the unit on electricity or magnetism. Each student receives a point for each correct placement of the word. An extra two points is received if the student can provide the definition and one point should be given if the student can correctly state it is falling under both electricity and magnetism (for example, electromagnetism). After the game, have students tally the points. You may want to provide prizes like a horseshoe magnet or fun refrigerator magnets.

Name _____

Magnets & Electricity Vocabulary Practice

Fill in the blank with the correct word from your electricity and magnetism vocabulary word sheet.

1. A _____ is a piece of iron or other metal that can be magnetized for a long time.

2. A _____ is material that easily allows an electrical charge to move through it.

3. A _____ slows the flow of electrons.

4. The unit of measure of current through a circuit is an _____.

5. A negatively charged particle around the nucleus of an atom is called an _____.

6. The movement of electrons is called _____.

7. A _____ acts like a "chemical pump" that forces electrons through a conductor.

8. A _____ is the path formed by the conductors for the electrons.

9. A suspended magnet with one end pointing toward Earth's magnetic North Pole is called a _____.

10. A _____ is any object that has a magnetic field.

11. Electrons that have built up in one place and are not moving create _____.

12. The center part of an atom is called the _____.

13. _____ is what keeps the protons and neutrons together.

14. A _____ happens when the electron path goes directly from the negative to the positive end of the battery without traveling through a resistor.

15. The positively charged particles of the nucleus are called the _____.

Name _____

Vocabulary Crossword Puzzle

Use the clues and the magnets and electricity vocabulary to fill in the crossword puzzle.

Across
1. negatively charged particle that travels around the nucleus of an atom
2. device that slows the flow of electrons
3. path formed by conductors for electrons
4. flow of electrons
5. physical property of matter in which there is an attraction due to unlike poles

Down
6. measure of how difficult it is to move electrons through a conductor
7. unit of measure of current through a circuit
8. collection and/or movement of electrons
9. this acts like a "pump" that forces electrons through a conductor
10. material that easily allows an electrical charge to move through
11. any object that has a magnetic field
12. a suspended magnet with one end pointing toward Earth's magnetic North Pole

Classroom Learning Centers

Just as the backdrops and costumes are important to a play, a welcoming classroom environment is important to foster learning. The room should be fun, inviting, and interactive. With that in mind, this section features learning center activities and bulletin board ideas to help you set up the room for magnets and electricity.

1. Interactive Web Site Learning Center

If you have enough computers, you could set up a small cluster of them. Direct students to various interactive web sites for children where they can ask questions, learn in-depth information about electricity and magnets, test their knowledge with quizzes, and watch video clips. Try this site to start: *http://www.brainpop.com* .

2. Static Electricity Learning Center

An electroscope allows students to test an object for a static charge. In this experiment, you'll have students test materials for static electricity using an electroscope. This is a perfect learning center after performing the balloon experiment in Lesson 1. You'll want to build the electroscopes yourself with the following materials unless you have a parent volunteer to help students operate the glue gun. You'll also need:
- several wide-mouthed clear plastic containers with plastic lids
- 2 small strips of holiday tinsel or aluminum foil for each container
- 2 common nails for each container
- balloons
- yarn
- cellophane wrap
- various objects and materials chosen by the children

Remove the lid from the container, and pierce it from the bottom with the sharp point of the nail. Then, glue the two strips of foil or tinsel to the head of the nail, and make sure they are in contact with the metal. Place the lid back onto the container so that the strips hang freely. Trim them if necessary. Students are now ready to test objects for an abundance of electrons (static electricity). When an object is brought near the point of the nail, the foil strips will separate to indicate the presence of a static charge. Have students record and compare the objects or materials that had a charge.

3. Designing Circuits Learning Center

This learning center gives students the opportunity to use the circuit material they experimented with during Lesson 4 to design and build their own circuits. This time, tell students to add switches and motors. After each circuit is made successfully, encourage students to draw it schematically and label it with the proper circuit name. Be sure to keep these materials at this learning center:
- D-size batteries and holder
- wires
- bulbs
- switches
- motors
- schematic drawings from Lesson 4

4. Experiment Learning Center

This center will help you organize all of the experiments in this book for your students. Be sure to have the following materials at this station:
- supplies for the experiments in **Step 6**
- the experiment pages for each lesson
- experiment **Science Logs**

You may also want to include directions for other experiments you've come across during your research.

Learning Centers Checklist: Teacher

Use this checklist to record which students have completed each center activity. Record a grade or a symbol to reflect the level of completion to the left of each learning center. You could use the wider column beneath each learning center to jot a note about the student's performance and the date completed.

Student		1. Interactive Web Site		2. Static Electricity Testing		3. Designing Circuits		4. Experiments

Learning Centers Checklist: Students

Photocopy this page for each student, and cut it in half. Have your students use this sheet to get sign-off by you or a peer each time they successfully complete a center. Remind students that completing more than one center a day or repeating a center during the week is permitted.

Name _____ Date_____

Centers Week ____ – ____	Monday	Tuesday	Wednesday	Thursday	Friday
1. Interactive Web Sites					
2. Static Electricity Testing					
3. Circuit Designs					
4. Experiments					

Name _____ Date_____

Centers Week ____ – ____	Monday	Tuesday	Wednesday	Thursday	Friday
1. Interactive Web Sites					
2. Static Electricity Testing					
3. Circuit Designs					
4. Experiments					

Bulletin Board Ideas

The bulletin board ideas below will help you and your students set up the room. Aside from these bulletin board ideas, you may want to put up a poster of the Periodic Table of Elements to refer to during the lesson as well as show a three-dimensional model of an atom. For magnets, you may find it useful to put up pictures of items that use magnetism to operate (such as a junkyard electromagnet) and include a poster showcasing the dos and don't when handling and storing magnets.

The Circuit Bulletin Board

Create an interactive bulletin board by covering strips of oak tag that measure 11 inches (27.9 cm) long and .5 inch (1.3 cm) wide with aluminum foil. Use pushpins to connect them together in many different arrangements. Then, place a battery pack on the board, and attach several light bulbs with holders around the board. Students can then challenge themselves to get as many bulbs to light as possible with wires and alligator clips. Be sure to explain the three different kinds of circuits with this bulletin board.

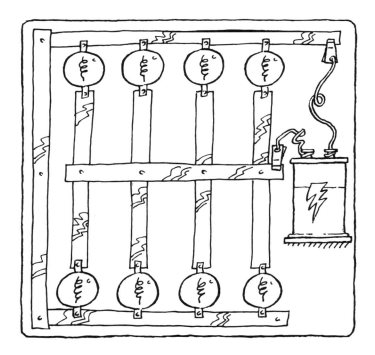

Magnet Mosaic

Hang a steel-backed whiteboard or use a portion of a large whiteboard to create an ever-changing bulletin board. Ask students to bring in the flexible refrigerator magnets their families no longer want and other magnets from around the house. Have students cut up the flexible magnets into various shapes and use the whiteboard to create a magnet mosaic. Take pictures of the various art displays, and hang them around the edges as the pieces get moved.

STEP 5
PLAN A PROJECT

Electricity and Magnetism Invention Quest

Requiring students to put their knowledge and skills to work is an ideal way to ensure long-term retention of content. In **Step 5: Plan a Project,** students have an opportunity to gather information over a long period of time and share their data. The end product is designs of inventions that utilize electric circuits, magnets, or both. Students can display these new inventions with an Electricity and Magnetism Invention Quest.

This project can be done individually but would also work well if students have a partner to help coordinate the design and building of the invention. You may want to start this project once students have a good understanding of the concepts in the electricity unit.

1. Gather Materials
Collect wires, batteries and holders, bulbs, motors, buzzers, and other low-voltage electrical appliances that will run on battery power. Include switches, too. Also collect a lot of magnets of different sizes and types. Ask for donations from parents and friends. You will need plenty of cardboard, glue, cloth, Styrofoam™, and other building materials. A checklist of materials appears on page 18. Next, think about who should be invited and begin getting invitations ready. Look to page 20 for a sample blank invitation.

2. Make a Design on Paper First
Decide if the students will be working individually on this project or in pairs. Then, give a sheet of large paper to each student or pair. Discuss the importance of sharing ideas and working with each other. Suggest that each student makes his or her own design first and then share elements from both designs. This approach often leads to better collaboration than if students try to design something together. Encourage students to be original and creative by asking them to think critically about their invention: *What will this invention do? Who or what will it serve? How will it be important in "real life?"* Students may want to use the **Electricity and Magnetism Invention Quest Checklist** on page 18 as a guide. Then, have students draw the final design on the paper. Remind them to use rulers, label the parts, and make a parts list. Verify all designs for feasibility before students proceed.

3. Time to Build
Allow students time to find and collect the materials they will need. Remind them to stay with their designs for now. Final adjustments can be made if the designs don't work. Try to set a deadline for the project, and send out your invitations now. Be sure you line up a space for the displays. Ask parent volunteers to be available for hot gluing and extra help.

4. Prepare Presentations
Discuss ways in which students will unveil their inventions to the public. Encourage the use of technology with digital photographs and software. Then, have students prepare a short description of the invention: its name, its uses, who or what it benefits, and its importance to the world. Once your students are ready, allow time for rehearsal of presentations.

5. Last-Minute Changes
This is a chance for students to make any last-minute adjustments or changes to the original design. Remind them to explain this step in their presentation.

6. Host an Invention Quest
Ask custodians to set up tables for you at the Invention Quest location. Be sure to have a spot, on a riser if possible, for each student or pair of students to show off their inventions and make presentations. Encourage peer review and evaluation.

Name _____

Electricity & Magnetism Invention Quest Checklist

Use the following checklist to mark off the items you used in your invention and the quantity of each.

Materials	Quantity		Materials	Quantity
❑ wire	_____		❑ bar magnet	_____
❑ battery	_____		❑ horseshoe magnet	_____
❑ battery holder	_____		❑ cardboard	_____
❑ bulb	_____		❑ glue	_____
❑ motor	_____		❑ cloth	_____
❑ buzzer	_____		❑ Styrofoam™	_____
❑ switch	_____		❑ Other: _____	_____
❑ round magnet	_____			

What my invention will do: _____

_____

_____

_____

_____

_____

_____

_____

_____

Why it is helpful: _____

_____

_____

_____

_____

_____

_____

_____

_____

Name _____

Weekly Invention Quest Goals

Use this chart to plan how you will complete your project. Budget your time for different parts of the project, from the design stage to drawing the design on paper, gathering materials, building your invention, making final adjustments, and making your final presentation to the public. Use this checklist to plan your activities and set goals.

Project Tasks & Goals for the week of ____ – ____	Monday	Tuesday	Wednesday	Thursday	Friday
Designing the Invention					
Drawing the Design					
Gathering Materials					
Building the Invention					
Making Final Adjustments					
Practicing the Presentation					
Other: _____					

You're Invited

Please Attend our Invention Quest

Come see student inventions that will change the world as we know it!

What? Electricity and Magnetism Invention Quest!

When?

Where?

What Time?

STEP 6

TEACH TEN TERRIFIC LESSONS

Introduction

The 10 lessons presented on the pages that follow provide a comprehensive study of magnets and electricity. Work through the steps in order, or pick and choose the activities that will enhance what you're already teaching — the choice is yours!

Each lesson contains 3 parts:

1. Teacher Note Page(s)

Provides a general overview of the lesson's topic. These pages include:

- **They'll Need to Know** ... for a general overview of the lesson's topic.
- **Prove It!** for points to bring up as students are working through the experiments.
- **Journal Prompt** to assess student learning and to give students the opportunity to put the science concept into their own words and/or expand their thinking beyond the topic.
- **Homework Idea** to follow up on the concept at home.

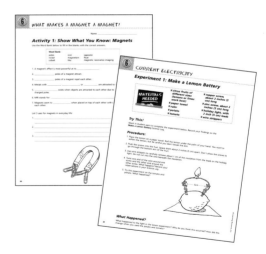

2. Experiments and Activities

Provides hands-on experiences designed to reinforce the day's lesson. The teaching notes page provides background information for each experiment.

3. Science Log

Provides a space for students to record the concepts learned and their observations from the experiments.

Overview

The following explains the objective of each lesson as well as the experiments, activities, and supplies needed in each lesson. Be sure to collect these supplies in advance.

Lesson	Supplies
1. Electrons: The ELECTR in Electricity Students learn about the atom and how electrons move around it.	**Experiment 1: Human Atom Model:** colored construction paper **Experiment 1: Static Electricity:** balloon, newspaper, aluminum foil, wool sweaters or socks, wax paper, cellophane wrap
2. Current Electricity Students learn about current electricity, how to make their own battery, and the principles of how a battery works.	**Experiment 1: Make a Lemon Battery:** citrus fruits, paper towels, rulers, potatoes, tomatoes, copper screws, zinc screws, string of holiday lights, wire strippers **Experiment 2: Make a Money Battery:** clean pennies, clean dimes, coffee filters, salt, hot water, copper wires, paper cups
3. Conductors vs. Insulators Students learn which materials conduct electricity and which do not.	**Experiment 1: Circuit:** D-size batteries, battery packs with leads, copper wires with alligator clips, thin iron wire (such as from picture frames) **Experiment 2: Lighting the Bulb:** flashlight bulbs, wires with alligator clips, D-size batteries **Experiment 3: Testing for Conductors and Insulators:** flashlight bulbs with bulb holders, D-size batteries with battery holders, wires with alligator clips **Activity 1: Show What You Know: Conductors:** page 40
4. Circuits Students learn about simple, series, and parallel circuits.	**Experiment 1: Simple and Series Circuits:** flashlight bulbs with holders, D-size batteries with holders and leads, wires with alligator clips **Experiment 2: Parallel Circuit:** flashlight bulbs with holders, D-size batteries with holders and leads, wires with alligator clips **Activity 1: Design and Test Your Own Circuits:** flashlight bulbs with holders, D-size batteries with holders, wires with alligator clips, page 43 **Activity 2: Show What You Know: Circuits:** page 45
5. Electrical Energy Sources Students use a debate exercise to explore and analyze the various resources that are used to produce electricity.	**Activity 1: Energy Sources Debate:** pages 47-49, computer with Internet access **Activity 2: Show What You Know: Energy Sources:** page 50

Overview *(continued)*

Lesson	Supplies
6. What Makes a Magnet a Magnet? Students learn about the characteristics of magnets by investigating how they work.	**Experiment 1: North and South:** 4-inch pieces of soft magnet about 1/8- to 1/4-inch thick, scissors **Experiment 2: Magic Floating Magnets:** donut magnets, Number 2 pencil, clay **Experiment 3: Floating Paper Clips:** strong magnets, paper clips, thread **Activity 1: Show What You Know: Magnets:** page 55
7. Magnetize/Demagnetize Students learn how to magnetize and demagnetize iron objects.	**Experiment 1: Magnetize a Nail:** bar magnets, 16D iron nails, small paper clips **Experiment 2: What Can Be Magnetized?:** bar magnets, 16D iron nails, small paper clips, other metallic materials and non-metallic materials
8. The Earth as a Magnet Students learn how a compass reacts to the Earth's magnetism and build their own compasses to test.	**Experiment 1: Test the Earth's Magnetism:** bar magnets, thread, access to a ceiling fixture **Experiment 2: Make a Compass:** sewing needles, cork, glass bowls, water, bar magnets
9. Electromagnets Students learn why an electromagnet works and how to make one.	**Experiment 1: Magnetic Attraction:** large batteries (6 volts or more), plastic or cardboard tubes, insulated bell wire, wire strippers, largest iron nails that will fit in tubes (sharp end dulled with a file), alligator clips **Experiment 2: Make a Basic Electromagnet:** 16D iron nails, insulated copper wire, wire strippers, alligator clips, D-size batteries with battery holders, paper clips **Experiment 3: Go Further with Electromagnets:** nails of various size and materials, insulated copper wire of various lengths, wire strippers, battery holders and D-size batteries, paper clips
10. Electric Motors Students learn how an electric motor works by building one of their own.	**Experiment 1: Move the Wire:** rulers, small disk magnets, 1.5-volt flashlight batteries, flexible wire, masking tape, wood boards, wire strippers or sandpaper **Experiment 2: Build An Electric Motor:** small disk magnets, large paper clips, plastic or paper cups, 20-gauge insulated copper wire, masking tape, 6-volt lantern batteries with holders, electrical lead wires with alligator clips at both ends, wire strippers, black permanent markers, page 68

Lesson 1: Electrons: The ELECTR in Electricity

Use this page when you introduce Electrons to your students. The fun facts can be used to draw your students into the topic.

They'll Need to Know ...

The atom is the smallest unit of matter, and it has unique characteristics. Atoms were once thought of as the smallest particles of matter, incapable of being split further, but we now know that they are made up of particles called protons, neutrons, and electrons.

Protons and neutrons make up the center of an atom. We call this center the nucleus. The nucleus of an atom has a certain number of protons and neutrons, depending on its position in the Periodic Table of Elements. The Table contains all hundred or so elements that have been discovered to date. The elements are those substances that make up all matter. The element Lithium, for example, has three protons and thus its atomic number is 3 (see Diagram A). Neutrons have no charge. Protons are positively charged and carry with them one unit of positive charge. They cling together with something called strong force. Usually, opposite charges attract, while like charges repel each other; however, the strong force is so powerful that protons cannot escape its hold.

Electrons are negatively charged. They orbit the nucleus randomly in what is called an electron cloud. Each electron maintains the same distance as it orbits, and its pathway is called a shell. Each electron carries one unit of negative charge. Some elements have more electrons than others depending on their position on the Periodic Table.

DIAGRAM A

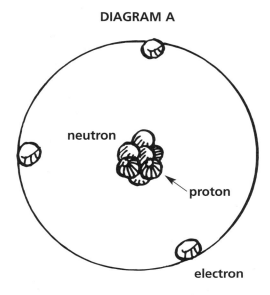

Lithium atom

• The atom was thought to be the smallest unit of matter until scientists discovered it had parts (electrons, protons, and neutrons) that were even smaller. We call these particles subatomic, or smaller than an atom.

Lesson 1: Electrons:
The ELECTR in Electricity (continued)

Elements that have the same number of protons and electrons are said to be neutral. When atoms in an element gain or lose electrons during a chemical reaction, they become electrically charged. Gaining electrons means that the atom has more electrons than protons. This causes it to carry a net negative charge, and it is called an ion (see the Fluorine ion in Diagram B). Losing electrons means the atom has more protons than electrons, which causes it to carry a net positive charge. Because electrons can move among atoms, they can build up in matter, a phenomenon known as static electricity.

DIAGRAM B

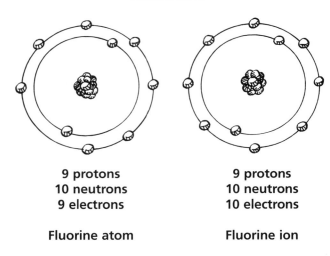

9 protons	9 protons
10 neutrons	10 neutrons
9 electrons	10 electrons
Fluorine atom	**Fluorine ion**

Check out this site to learn more about the Periodic Table of Elements, each atom's atomic weight, its position on the table, its abbreviation, and its structure: *http://www.webelements.com/*

Prove It!

In a whole group setting, define and discuss atoms and their structure using the board to illustrate your points. Work as a class through the first experiment. Allow time for students to take turns. Distribute a **Science Log** handout to each student after the first experiment, and have them fill it out. Then, allow students to work in pairs for the second experiment. You'll want to go through the teaching notes with students so they understand the concepts after they've worked through the experiment.

Experiment 1: Human Atom Model Teaching Notes:
In this experiment, students model electrons and how they are constantly moving. This fact is critical to understanding electricity. Be sure to allow each student to role-play as an electron, proton, and neutron.

Experiment 2: Static Electricity Teaching Notes:
In this activity, students discover which materials give up electrons. When they rubbed the balloon on their hair, their hair gave up electrons to the balloon. When the balloon was brought close to the paper pieces (which are positively charged due to their make-up) they jumped up to the balloon because of the opposite attraction. As each of the other materials (aluminum foil, wool sweater or sock, wax paper, and cellophane wrap) was rubbed against the balloon, some gave up electrons and some did not, as evidenced by the reaction of the paper pieces.

Journal Prompt

Using your knowledge of static electricity, explain what might be happening when lightning strikes.

Homework Ideas

Challenge students to test different materials for creating static electricity, such as different floor surfaces and footwear. Have students answer the following questions on their own: *Which ones work the best? Which don't seem to work at all?*

Experiment 1: Human Atom Model

MATERIALS NEEDED

- **3 sheets each of 2 different colors of construction paper**
- **large, open area**
- **5 sheets of another color of construction paper**

Try This!

Work as a class to complete the experiment listed below. Record your findings on the **Human Atom Model** Science Log.

Procedure:

1. Fold and tear the sheets of construction paper in half. You should have six half-sheets of one color, six half-sheets of another color, and ten half-sheets of a third color.

2. Label five same-colored sheets of paper with the word "proton." Draw a plus sign below the word.

3. Label five same-colored sheets of paper with the word "neutron." Draw a zero symbol under the word.

4. Label the 10 same-colored sheets of paper with the word "electron." Draw a minus sign below the word.

5. Five students should hold a proton sheet, and five students should hold a neutron sheet. All these students should then gather very closely in the center of an open area. Together, they represent the "nucleus."

6. Students should hold the remaining electron cards and stand at different distances from the nucleus group.

7. The students role-playing the electrons should circle about the nucleus at random speeds and in random directions while staying the same distance away.

8. Have students take turns playing different parts.

What Happened?

What were you doing in this experiment? What charge were you? Were the electrons always moving?

Experiment 2: Static Electricity

- **balloon**
- **clean head of hair**
- **tiny bits of torn newspaper**
- **sheet of aluminum foil, about 1 foot long (.3 m)**
- **wool sweater or sock**
- **sheet of wax paper, about 1 foot long (.3 m)**
- **sheet of cellophane wrap, about 1 foot long (.3 m)**

Try This!

Work with a partner to complete the experiment listed below. Record your findings on the **Static Electricity** Science Log.

Procedure:

1. Blow up the balloon. Rub it against your hair. What do you notice?

2. Sprinkle the bits of newspaper on a table.

3. Hold the balloon near the paper pieces, but don't let it touch them. What happens? Record your observations in your Science Log.

4. Rub the balloon with the aluminum foil. Now try Step 3 again.

5. Repeat Step 3 by rubbing the balloon with each of the other materials listed above. Record your results in your Science Log.

What Happened?

What happened when you pulled the balloon away from your hair? What does this mean? What did the balloon do when it was close to the paper? How did the balloon react when rubbed with the other materials? Since positive and negative charges attract, do you think this theory was at play in this experiment? Explain your thoughts to your partner.

Name _____

Science Log

Use this sheet to record what you observed during **Human Atom Model** experiment.

Question: How do electrons move about in an atom's nucleus?

_____

_____

What parts did I play?

_____

_____

What did I do for each part?

_____

_____

Why does each part behave this way?

_____

_____

Draw a picture of an atom and its particles. Label the nucleus, neutrons, protons, and electrons.

Name _____

Science Log

Use this sheet to record what you observed during your **Static Electricity** experiment.

Question: Why did the newspaper pieces cling to the balloon at some times, but not others?

_____

_____

_____

What I did:

_____

_____

_____

What happened:

_____

_____

_____

Why it happened:

_____

_____

_____

Lesson 2: Current Electricity

Use this page when you introduce Current Electricity to your students. The fun facts can be used to draw your students into the topic.

They'll Need to Know ...

A battery is essentially a can of chemicals that stores and produces electrons. The chemical reaction only takes place when the battery is in operation. An operating battery is like a pump that pushes electrons through wires or conductors.

A battery stores a certain amount of energy. The unit to describe this stored energy is called a volt. You will be working with 1.5-volt batteries. This means the battery is capable of producing 1.5 volts of electricity at any one time. This is also known as potential energy.

Your 1.5- volt battery contains a paste that helps to create the chemical reaction necessary for it to operate. Other batteries, such as a car battery, contain a liquid electrolyte or acid that helps to create the chemical reaction.

A battery has a positive (+) end, or cathode, and a negative (–) end, or anode. Electrons collect on the negative side of the battery and flow to the positive side when connected to a complete circuit. Connecting a wire from the negative to the positive end will cause the electrons to flow there quickly. This creates the current called current electricity. Current can be measured in a unit known as an ampere.

The kind of current that comes from a battery is called direct current because it moves in only one direction, from the negative end to the positive end of the battery. An alternating current comes from an outlet in your home. This current flows in two directions.

Prove It!

In a whole group setting, review the steps involved in the **Making a Lemon Battery** experiment. Divide students into pairs or small groups to complete the experiment. Be sure to demonstrate how to use a wire stripper before you begin the experiment. As each group finishes, distribute other types of fruit and vegetables (potatoes work well) so that students may try to make other batteries.

Experiment 1: Make a Lemon Battery Teaching Notes:
In this experiment, zinc and copper react in the juice of a vegetable or fruit to produce electricity. The lemon does not participate directly in the reaction. It is there to help transport the zinc and copper ions in the solution while keeping the copper and zinc apart. If the zinc and copper are in contact with each other, only heat will be generated. By keeping them apart, the electron transfers over the wires of the circuit to light the bulb.

Experiment 2: Make a Money Battery Teaching Notes:
In this experiment, zinc and copper react again, this time in a salt solution to produce a small electric current. The zinc in the dimes reacted with the copper in the pennies and used the salt solution as an electrolyte to create a small electric current. This experiment takes currents one step further by introducing metals. Be sure to explain the role that copper and zinc have here.

Journal Prompt

Write about what it would be like if there were no batteries. How would your life change? What could you use instead of batteries that would be practical?

Homework Idea

Have students try this experiment at home after conducting **Experiment 1**: Compare the brightness of the light to fruit size and type. What can you conclude about why the light may have been brighter, dimmer, or did not change at all? Try the same experiment with:
1) different-size screws, and 2) same-size fruit. Compare the results.

FUN FACTS!

• The first electric battery was called the "voltaic pile."

Experiment 1: Make a Lemon Battery

- **citrus fruits of different sizes (lemons or limes work best)**
- **paper towel**
- **ruler**
- **potato**
- **tomato**
- **copper screw, about 2 inches (5 cm) long**
- **zinc screw, about 2 inches (5 cm) long**
- **holiday light, with 2 inch (5 cm) leads**
- **wire strippers**

Try This!

Work in student pairs to complete the experiment below. Record your findings on the **Make a Lemon Battery** Science Log.

Procedure:

1. Place the lemon on a paper towel. Roll the lemon under the palm of your hand. You want to soften the lemon, but be careful you don't break the skin.

2. Push the screws into the fruit. Space them about 2 inches (5 cm) apart. Don't allow the screws to go through the bottom skin of the fruit.

3. Use wire strippers to carefully remove about 1 cm of the insulation from the holiday light leads. Do not cut into the wire beneath the insulation.

4. Twist one end of the wire around one screw and the other end around the other screw. What happens? Record your observations in your Science Log.

5. Try this experiment on the tomato and potato. What happened?

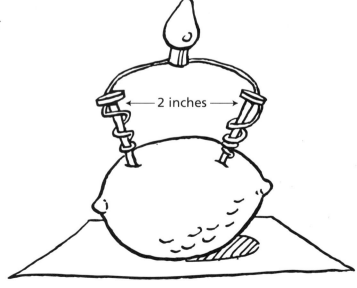

What Happened?

What happened to the light in the lemon experiment? Why do you think this occurred? How did this change when you used the potato and tomato?

Experiment 2: Make a Money Battery

MATERIALS NEEDED

- 6 clean pennies
- 6 clean dimes
- coffee filter paper, cut into 6 pieces that measure 1 square inch
- cup of salt
- quart of hot water
- 2 4-inch pieces of copper wire with 2 cm of coating stripped off ends
- paper cup

Try This!

Work in small groups to complete the experiment. Record your findings on the **Make a Money Battery** Science Log.

Procedure:

1. Mix the salt and water together in a paper cup, stirring to dissolve the salt.

2. Soak the pieces of filter paper in the saltwater solution for about one minute.

3. Place one dime on the table. Cover it with one piece of filter paper.

4. Place a penny on top of the filter paper. Place a dime on top of the penny.

5. Cover the two coins with another piece of filter paper.

6. Continue stacking layers of pennies, dimes, and paper until all the dimes are used.

7. Place the final piece of filter paper over the top dime. Place the last penny on top.

8. Carefully pick up the stack of dimes, pennies, and filter paper. Hold it between your thumb and pointer finger.

9. Have a partner slide a wire between your pointer finger and the penny. Slide another wire between your thumb and the dime.

10. Touch the ends of the wires to your tongue. What happens? Record your observations in your Science Log.

11. Have your partner try the experiment with a new wire.

What Happened?

What did you feel when you pressed the wire to your tongue? What do you think happened? Why?

Name _____

Science Log

Use this log sheet to record your findings from the **Make a Lemon Battery** experiment.

Question: Are the materials that make up the screws important in this experiment? Why?

_____

_____

What I did:

_____

_____

What happened:

_____

_____

Why it happened:

_____

_____

What do you think you could use besides screws in this experiment?

_____

_____

Fruit and Vegetable Battery Testing Chart	
Type of Fruit or Vegetable	Did the Bulb Light? (Yes or No)

Name _____

Science Log

Use this log sheet to record the findings from the **Make a Money Battery** experiment.

Question: What do you think would happen if you used quarters instead of dimes?

_____

_____

_____

What I did:

_____

_____

_____

What happened:

_____

_____

_____

Why it happened:

_____

_____

_____

What are some other objects made out of zinc or coated with zinc that may work with this experiment?

_____

_____

_____

Lesson 3: Conductors vs. Insulators

Use this page when you introduce Conductors vs. Insulators to your students. The fun facts can be used to draw your students into the topic.

They'll Need to Know ...

Charged particles flow most easily through conductors, such as metals or some liquids like saltwater. Electrons in metals are loosely attached to the atoms, so they can move easily. The human body (which is mostly saltwater) is also a good conductor, which is why electric shocks can be so dangerous. Insulators like rubber, wood, glass, and plastics do not conduct electricity well. Their electrons are tightly bound to their atoms and do not move easily.

Electricity will only flow when a power source, such as a battery or a generator, sets the electrons in full circle. For example, electrons flow from a battery down a wire to a light bulb, through the filament of the bulb, then back up another wire to the battery. This closed loop is called a circuit.

When a light bulb comes into contact with the poles of the battery and electrons pass through the filament of a bulb, there is great resistance. This resistance causes extreme heat and the filament glows. That's how light is produced.

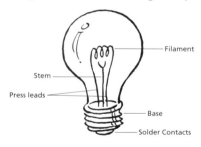

Filament
Stem
Press leads
Base
Solder Contacts

Electrical energy always seeks the route with the least resistance around a circuit back to the source, like a battery. If the wires in a circuit both touch a conductor, such as a metal tabletop, the electrons will take the shorter route (across the tabletop) back to the battery, rather than travel to the light bulb. This is called a short circuit. If the two poles of a battery are directly connected it causes a short circuit to heat up the battery and possibly cause a fire.

FUN FACTS!

• A lightning bolt can contain 3750 kilowatts of electrical energy.

Prove It!

Discuss that electricity always takes the shortest path to complete the circuit. If that shortest path connects both ends of the battery without first going through a resistor, the result is a short circuit, which can quickly shorten the life of a battery.

Experiment 1: Circuit Teaching Notes:
In this experiment, the copper alligator wire became a very good conductor with very little resistance. When students added the iron wire, they probably saw it was not as good a conductor. The higher resistance caused friction from the electrons traveling through the wire. This resulted in heat.

Experiment 2: Lighting the Bulb Teaching Notes:
In this experiment, there are four ways to light the bulb. They are shown below. As referenced, as long as students made a complete path for the electrons to travel through the bulb, the bulb will light up.

Experiment 3: Testing for Conductors and Insulators Teaching Notes:
As students touched the wire ends to conductors, they completed the circuit needed to light the bulb. When they touched an insulator, the circuit was not completed and the bulb did not light.

Journal Prompt

If water is a conductor and we are made mostly of water, why is it important for us to be very safe with electricity?

Homework Idea

Have students think about how electricity travels through wires to get to our homes, schools, and businesses every day. What would happen if we suddenly did not have insulators to help us? What alternative could there be to get our electricity?

CONDUCTORS VS. INSULATORS

Experiment 1: Circuit

- **2 D-size batteries**
- **2 copper wires with alligator clips on both ends**
- **piece of thin iron wire**
- **battery pack with leads**

Try This!

Work in small groups to complete each of the experiments listed below. Record your observations on the corresponding Science Logs.

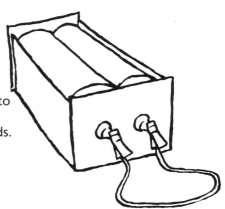

Procedure:

1. Place the batteries in the holder. Connect one end of each copper wire to one end of each battery lead using the alligator clips.
2. Observe and feel what happens to the iron wire after only a few seconds.
3. Disconnect the copper wires from the battery **immediately**.
4. Repeat Steps 1-3 with the iron wire.

What Happened?

What happened when the copper wire was connected to the battery wire? How about when you replaced the copper wire with the iron wire? What does this experiment show about copper vs. iron as conductors?

Experiment 2: Lighting the Bulb

- **flashlight bulb**
- **1 wire, with about 2 cm insulation**
- **stripped from each end**
- **alligator clips**
- **D-size battery**

Procedure:

1. Use the objects listed above to try to light the bulb. There are four ways to do this. Draw a picture to the right for one successful method you design.

What Happened?

Were you able to light the bulb using all four ways? What were some similarities between your methods?

Experiment 3: Testing for Conductors and Insulators

MATERIALS NEEDED

- **flashlight bulb with bulb holder**
- **3 wires with alligator clips**
- **battery holder with 2 D-size batteries**

Try This!

Work in a group to complete the experiment below. Record your findings on your **Testing for Conductors and Insulators** Science Log.

Procedure:

1. Connect the alligator clip holding one end of a wire to the cathode (+) lead of the battery holder. Connect the alligator clip at the other end of the wire to the base beneath the flashlight bulb.

2. Connect another wire to the anode (-) lead on the battery holder. Do not connect this wire to the battery holder.

3. Connect one end of the third wire to the other end of the bulb holder.

4. You should have two wire ends free, one that's connected to the battery and one that is connected to the bulb holder. Find objects around the room to touch the two wire ends to. If the bulb lights up, then the object is a conductor. If the bulb does not light up, then the object is an insulator.

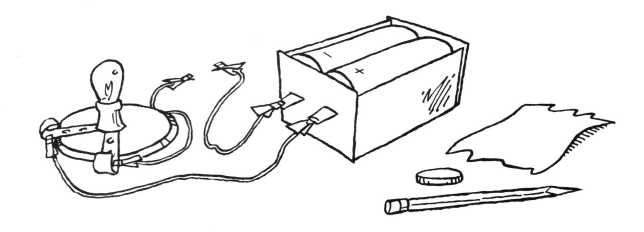

What Happened?

Did the bulb always light up when you touched the wires to each object? What objects caused the bulb to light up? Does that make them conductors or insulators? What objects didn't light up the bulb? What does that make them?

Name _____

Science Log

Use this log sheet to record the findings from your experiments.

Circuit

Question: What causes a short circuit?

_____

What I did: _____

_____

What happened: _____

_____

Why it happened: _____

_____

Lighting the Bulb

Question: Why do you think there is glass covering the filament of the bulb?

_____

What I did:

_____

What happened?_____

_____

Why it happened: _____

_____

_____

Name _____

Science Log

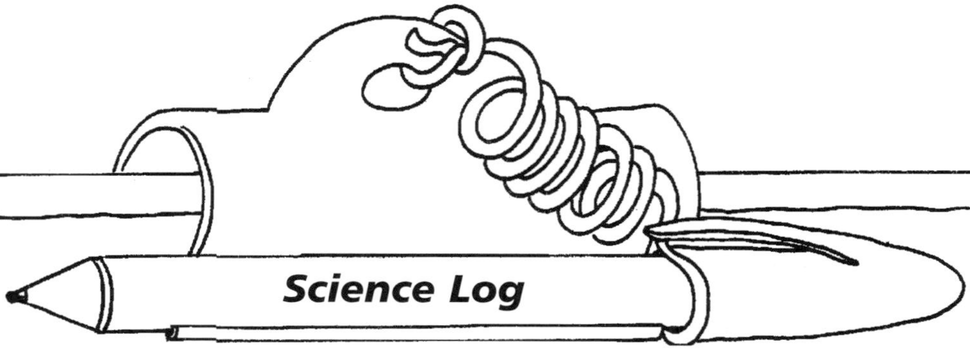

Use this log sheet to record the findings from **Testing for Conductors and Insulators** experiment.

Question: What materials make the best conductors? _____

_____

_____

_____

Record your test results here. Include the names of the insulators and conductors.	
Test Results Table	
Insulator (bulb did not light)	Conductor (bulb lit up)

What I did: _____

_____

_____

_____

Name _____

Activity 1: Show What You Know: Conductors

Draw a closed loop circuit of electricity flowing to a light bulb. Be sure to label all the parts: light bulb with holder, battery holder with batteries, and wires.

Lesson 4: Circuits

Use this page when you introduce Circuits to your students. The fun facts can be used to draw your students into the topic.

They'll Need to Know ...

A complete circuit must begin at the negative end of the energy source and end at its positive end. The examples below are different types of circuits.

- The three components of a circuit are an energy source, a conductor, and a receiver or resistor.
- In a simple circuit (A), there is only one resistor (such as the bulb shown in this example).
- In a series circuit (B), there are two or more resistors connected to each other.
- In a parallel circuit (C), there are two or more resistors, but they are connected to the battery, not to each other.

A circuit is often represented by a schematic diagram (such as those shown below.) The symbols that represent each part of a circuit in a schematic diagram are also shown.

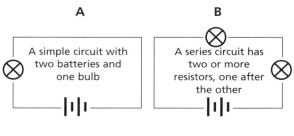

A
A simple circuit with two batteries and one bulb

B
A series circuit has two or more resistors, one after the other

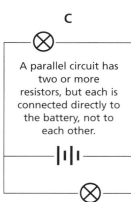

C
A parallel circuit has two or more resistors, but each is connected directly to the battery, not to each other.

battery   wire   bulb

Prove It!

Draw and label a simple, series, and parallel circuit on the board using bulbs as the resistors. Also, draw the schematic symbols for student reference. Explain the importance of using a ruler when drawing the wires and the fact that schematic drawings don't look like the real thing. Then, review the directions for each experiment with your students.

Experiment 1: Simple and Series Circuits Teaching Notes:
In this experiment, students lit a bulb by making a complete circuit. However, when students connected two bulbs in a series circuit, the bulbs had to share the electrons, so they both appeared dimmer.

Experiment 2: Parallel Circuit
In a parallel circuit, the electrons flow in two separate paths. This experiment on parallel circuits has students provide both bulbs with equal electric power. Be sure to point this out, and ask students whether both bulbs give off the same amount of light.

Journal Prompt

Just one flick of a switch can turn on many lights in a hallway. How can that be, and why are they all equally bright? What would make them dimmer?

Homework Idea

Challenge students to write a short story that tells the trip an electron makes as its travels through a circuit.

FUN FACTS!

- Thomas Edison invented the light bulb in 1879.

Experiment 1: Simple and Series Circuits

MATERIALS NEEDED

- *flashlight bulb and holder*
- *2 wires with alligator clips*
- *2 D-size batteries with holder and leads*

Try This!

Work in small groups to complete the experiment below. Record your findings on your **Simple and Series Circuits** Science Log.

Procedure:

1. Connect one end of each wire to each side of the bulb holder.
2. Connect the other ends of the wires to the battery holder leads. This is a simple circuit.
3. If the bulb does not light up, check the connections and try new wires or a new bulb.
4. Draw a schematic of this circuit in your Science Log.
5. Repeat the activity with two bulbs, connected by one more wire.

What Happened?

What happened during the simple circuit? How about when you created a series circuit?

Experiment 2: Parallel Circuit

Try This!

Follow the directions below to complete the experiment. Record your findings on your **Parallel Circuit** Science Log.

MATERIALS NEEDED

- *2 flashlight bulbs and holders*
- *4 wires with alligator clips*
- *2 D-size batteries with holders and leads*

Procedure:

1. Make a simple circuit like you did in the previous activity.
2. Now add another bulb holder to the circuit so two wires are connected to each battery lead instead of one. Try different ways to make these connections and keep the bulbs brightly lit.
3. Draw the circuit in your Science Log as you complete it.

What Happened?

What happened to both bulbs in this experiment? Why do you think that occurred?

CIRCUITS

Activity 1: Design and Test Your Own Circuits

Use the materials from **Experiments 1 & 2** to create your own circuits. Try to combine circuits to build something creative.

• **materials from previous experiments, with additional wires and bulbs**

Procedure:

Follow the same steps as before. Add one type of circuit at a time or string many together. Keep track of bulb brightness. Record your results below:

What I did: _____

_____

_____

_____

What happened
(draw schematics here):

Why it happened: _____

_____

_____

Name _____

Science Log

Use this log sheet to record the findings from your circuit test experiments.

Question: Compare the three different kinds of circuits (simple, series, and parallel). How are they alike and how are they different?

_____

_____

Simple and Series Circuits

What I did
(draw schematics here):

What happened: _____

_____

Why it happened: _____

_____

Parallel Circuit

What I did
(draw schematics here):

What happened: _____

_____

Why it happened: _____

_____

Activity 2: Show What You Know: Circuits

Draw examples of the following types of circuits: **simple**, **series**, and **parallel**.

Use electrical schematic symbols in your drawings.

simple circuit

series circuit

parallel circuit

Lesson 5: Electrical Energy Sources

Use this page when you introduce Electrical Energy Sources to your students. The fun facts can be used to draw your students into the topic.

They'll Need to Know ...

Electricity is delivered by a power company for use in our homes. The process begins as electricity from power plants is sent through a transformer and is boosted to a very high voltage. The high-voltage power lines carry electricity into cities and towns, where it is delivered to substations and to transformers that lower the voltage. This lower-voltage power then passes into a smaller transformer near our homes, where power is reduced even further. It is carried above ground or below ground into our houses. It must pass through a meter so the electric company can bill us for the power that we use.

It is important for citizens to know where electricity comes from, because we rely on it for so many aspects of our lives. The growing need for energy has put a strain on our ability to produce it, as past brownouts in California have proven. Researching the variety of electricity sources will help students begin to understand the enormous task of bringing electricity into their lives.

Prove It!

The activity in this unit has students organize a debate on the pros of a selected energy source. A debate is an organized argument. Each team gathers facts on the assigned topic. (Remind students that they may have to defend a point of view with which they might not agree.) After facts are gathered, student teams create convincing statements to support their point of view.

As a team presents its argument, its members should read statements clearly, accurately, and with authority. (You will need to model this and provide students the opportunity to practice this beforehand.) Each student takes notes on the other teams' facts and jots down questions about any points that seem unclear. These questions can then be asked during a brief rebuttal period after a team has finished its presentation.

Discuss with your students what a debate entails and the best way to gather information to support a topic. Facts work best for this age. Have students organize facts into categories. Students should then practice using their facts to support their topic. You may even wish to have student teams write a statement that incorporates many of the facts about their energy source.

Students will need to be aware of the following debate guidelines:

1. Presentations cannot exceed five minutes.
2. Opposing teams have 30 seconds to ask each rebuttal question.
3. The defending team has one minute to answer each rebuttal question.
4. Each team can ask no more than three rebuttal questions.

One helpful web site to get students started, and which has information for each team, is:

http://www.energy.ca.gov/education/story/story-html/story.html#table

Journal Prompt

Think of how dependent we are on electrical energy. In some parts of the world, however, people don't have electricity. Discuss some ways your life is different here from people who live in the jungle where there is no electricity.

Homework Idea

After learning about the different sources of electricity, have students select one that they feel best benefits the Earth and its human population. Have them explain why they chose this source and which source they feel is the most harmful for the Earth.

- The surface of the Sun is about 10,000 degrees Fahrenheit. The center is about 25 million degrees Fahrenheit. No wonder it is a great source for solar power!

Activity 1: Energy Sources Debate

Once you have been assigned your topic, explore the following web sites or books for information.

MATERIALS NEEDED

- **Geothermal Energy:**
  http://www.energy.ca.gov/education/
  story/story-html/chapter04.html

- **Fossil fuels (coal, oil, and natural gas):**
  http://www.energy.ca.gov/education/
  story/story-html/chapter05.html

- **Hydroelectric Power:**
  http://www.energy.ca.gov/education/
  story/story-html/chapter06.html

- **Nuclear Energy:**
  http://www.energy.ca.gov/education/
  story/story-html/chapter07.html

- **Ocean Energy:**
  http://www.energy.ca.gov/education/
  story/story-html/chapter08.html

- **Wind Energy:**
  http://www.energy.ca.gov/education/
  story/story-html/chapter10.html

- **Solar Energy:**
  http://www.energy.ca.gov/education/
  story/story-html/chapter09.html

- **Done in the Sun: Solar Projects for Children**
  by Astrid Hillerman, Anne Hillerman, and Mina Yamashita

- **Water and Wind Power**
  by Martin Watts

- **Sun Power: Facts about Solar Energy**
  by Steve J. Gadler and Wendy Adamson

- **Atom at Work: How Nuclear Power Can Benefit Man**
  by C. B. Colby

- **Earth Power: The Story of Geothermal Energy**
  by Madeleine Yates

- **Energy From Wind and Water**
  by Donna Bailey

Procedure:

1. You will be assigned a source for producing electricity. Your job is to make a statement such as "Solar energy is the best method for producing energy because …" Then, you will need to find facts that support why this statement is true. Use the **Energy Sources Debate Worksheet** for help.

2. Be sure your facts are accurate and clearly support your position.

3. When all groups are ready, you will have a debate over which method for producing electricity is best for us and the Earth. Review the debate rules with your teacher.

4. As teams present their arguments, take notes about different energy sources. Use the **Record Sheet.** Prepare questions about points you feel were not clearly made.

Name _____

Energy Sources Debate Worksheet

Use this worksheet to help you prepare for the debate.

Energy source I'm responsible for:

_____

My energy source is good for the environment because:

_____

_____

_____

Facts that support this statement include:

_____

_____

_____

My energy source is good for people because:

_____

_____

_____

Facts that support this statement include:

_____

_____

_____

_____

_____

_____

Name _____

Record Sheet

Source of Energy	Supporting Facts	Questions

ELECTRICAL ENERGY SOURCES

Name _____

Activity 2: Show What You Know: Energy Sources

Use the space provided below to write an essay discussing what you feel to be the best energy source for the future.

_____

_____

_____

_____

_____

_____

_____

_____

_____

_____

_____

_____

_____

_____

_____

_____

Lesson 6: What Makes a Magnet a Magnet?

Use this page when you introduce Magnets to your students. The fun facts can be used to draw your students into the topic.

They'll Need to Know ...

- Magnets can be temporary, permanent, or natural.
- Magnets have a north pole and a south pole.
- Magnets can be flexible or rigid.
- All magnets have an invisible magnetic field around them (see illustration below).
- A magnet is most powerful at its poles.
- Opposite poles of a magnet attract while like poles repel each other.
- Metals with iron, nickel, or cobalt are attracted to magnets.
- Magnetism exists when objects are attracted to each other due to their oppositely charged poles.
- When magnets are placed on top of each other with like poles facing each other, they seem to float.
- Magnets get their name from a naturally occurring magnetic ore called magnetite.

It is important to know about magnets and their characteristics because they play a very important part in our lives. Every electric motor, every computer, and just about every other electrical device has a magnet in it somewhere. Magnets are used in medicine, especially with an MRI (Magnetic Resonance Imaging) machine, which gives doctors a clearer image than an x-ray. Magnets even help with transportation in subway trains and in the newer levitating trains.

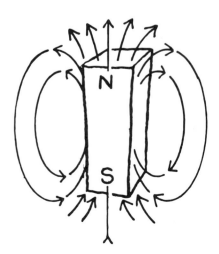

Prove It!

Discuss the directions for the experiments in this lesson with students. Be sure to instruct students to take their time in the **Floating paper clips** experiment. It may take several times before it works properly.

Experiment 1: North and South Teaching Notes:
Every magnet, no matter how small, has a north pole and a south pole. In this experiment, the cut ends of the magnet became poles opposite to the other end. Students probably noted that when they lined up the cut ends of the magnet, the same poles touched, so the magnet pieces repelled each other. Then, when they lined up a cut end with an uncut end of the magnet, opposite poles touched, so there was enough attracting force for the pieces to stick together.

Experiment 2: Magic Floating Magnets Teaching Notes:
In this experiment, the first three magnets combined to make one strong magnet. The other three magnets were forced away by the strong magnetic field. This made the top three magnets appear to float.

Experiment 3 Floating Paper Clips Teaching Notes:
In this experiment, students will notice that the strong magnetic field was still acting on the paper clip even though it was not touching the magnet.

Journal Prompt

Make a list of all the uses for a magnet.

Homework Idea

Magnetite and lodestone are naturally occurring magnetic materials. Have students find out more about them from an encyclopedia or the Internet.

- Some people use a magnetic burglar alarm to protect their cars and houses.

WHAT MAKES A MAGNET A MAGNET?

Experiment 1: North and South

 **MATERIALS NEEDED**
- **3 4-inch long pieces of soft magnet 1/8-inch thick to 1/4-inch thick**
- **scissors**

Try This!

Work in small groups to complete the experiment below. Record your findings on the **North and South** Science Log.

Procedure:

1. Cut the strip of magnet in half. Press the cut ends together. What happens?

2. Turn one of the pieces around. Press the cut end of one piece against the uncut end of the other piece. What happens? Record your observations in your Science Log.

What Happened?

What happened when the two magnets met the first time? How about the second time? Can you explain why this occurred?

Experiment 2: Magic Floating Magnets

Try This!

Work in small groups to complete the experiment below. Record your findings on the **Magic Floating Magnets** Science Log.

Procedure:

1. Push the pencil into the ball of clay. Don't push it all the way through. The clay will serve as a stand.

2. Slide three of the magnets onto the pencil. Be sure their poles are arranged north to south. They should stick together, or attract each other.

3. Place the next magnet on the pencil. Line up its north pole with the north pole on the magnet below it. Add the last two magnets arranging them the same way as you did with the first three. What happens? Record your results in your Science Log.

• **6 donut magnets**
• **Number 2 pencil**
• **Ball of clay**

What Happened?

What happened to the first three magnets? What happened when you slid a fourth magnet on the pencil? Why did this occur?

Experiment 3: Floating Paper Clips

Try This!

Work in small groups to complete the experiment below. Record your findings on your **Floating Paper Clips** Science Log Sheet.

Procedure:

1. Tie a 10-inch (25-cm) piece of thread to one end of a paper clip.
2. Tie the other end of the thread to your desk.
3. Pick up the paper clip with the magnet.
4. Use the magnet to slowly pull the paper clip until the thread is tight.
5. Keep pulling the magnet away slowly until the paper clip is no longer touching the magnet.

• **strong magnet**
• **paper clips**
• **thread**

What Happened?

What happened to the paper clip when the magnet was next to it? What happened to the paper clip when the magnet was nearby, but not touching it? Why do you think that occurred?

Name _____

Science Log

Use this log sheet to record the findings from your magnet experiments.

Question: What is a magnetic field? _____

_____

North and South ·

What I did: _____

_____

What happened: _____

_____

Why it happened: _____

_____

Magic Floating Magnets ·

What I did: _____

_____

What happened: _____

_____

Why it happened: _____

_____

Floating Paper Clips ·

What I did: _____

_____

What happened: _____

_____

Why it happened: _____

_____

Name _____

Activity 1: Show What You Know: Magnets

Use the Word Bank below to fill in the blanks with the correct answers.

> **Word Bank:**
>
> poles   iron   opposite
> nickel   magnetism   float
> cobalt   like   magnetic resonance imaging

1. A magnet's effect is most powerful at its _____.

2. _____ poles of a magnet attract.

3. _____ poles of a magnet repel each other.

4. Metals with _____, _____, or _____ are attracted to magnets.

5. _____ exists when objects are attracted to each other due to their oppositely charged poles.

6. MRI stands for: _____.

7. Magnets seem to _____ when placed on top of each other with same poles facing each other.

8. List 3 uses for magnets in everyday life:

1. _____

   _____

2. _____

   _____

3. _____

   _____

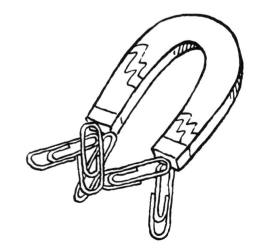

Lesson 7: Magnetize/Demagnetize

Use this page when you introduce the concepts of Magnetizing and Demagnetizing to your students. The fun facts can be used to draw your students into the topic.

They'll Need to Know ...

When the atoms inside an object are lined up according to their poles, a magnetic field is created. In nonmagnetic materials, the atoms are arranged in a random order. When a magnetic field comes in contact with a nonmagnetic material, such as an iron nail, the atoms in the iron nail line up, creating a small magnetic field. The iron nail is then attracted to the magnet. Sometimes the iron nail's atoms remain lined up for a short time, creating a temporary magnet. This weak magnet can be demagnetized by striking it on a hard surface. This jumbles the atoms once more.

Prove It!

Review the directions for the following experiments. Remind students that the temporary magnet is just that: temporary. In some cases the magnetic field only lasts a few seconds, so students must act quickly to pick up paper clips.

Experiment 1: Magnetize a Nail Teaching Notes:
In this experiment, be sure to point out that the nail itself does not have a magnetic field. But, when students held the magnet to the nail, it created a magnetic field around the nail making it an extension of the magnet. Then, by stroking the nail on the magnet they lined up the atoms temporarily in the nail, which caused it to become a temporary magnet. Striking the nail on a hard surface demagnetized it by destroying the orderly arrangement of atoms in the nail.

Experiment 2: What Can Be Magnetized? Teaching Notes:
In this experiment, students try to create temporary magnets from other objects. Remind students of the metals that can create magnetic fields upon completion of this experiment (i.e., iron, cobalt, and nickel).

Journal Prompt

Explain the difference between a permanent magnet and a temporary magnet. Which is more useful and why?

Homework Idea

Ask students to interview their parents to find out what they think are important uses for magnets. Have them list their parents' responses and write about whether they agree.

- Neodymium-Iron-Boron (NIB) supermagnets are so strong that eye protection must be worn when working with them. This is because if they slam together, they can shatter and eject chips at dangerous speeds.

Experiment 1: Magnetize a Nail

Try This!

Work in small groups to complete the experiment below. Record your findings on your **Magnetize a Nail** Science Log.

MATERIALS NEEDED

- *bar magnet at least 3 inches (7.6 cm) long*
- *16D iron nail*
- *small paper clips*

Procedure:

1. Place 10 or so small paper clips in a pile on a desk.

2. Try to pick up a paper clip with the nail. What happens?

3. Hold the nail against the magnet. Try to pick up paper clips again using this combined tool. What happens?

4. Slowly stroke one end of the nail against one end of the magnet at least 20 times. Be sure to stroke in the same direction each time.

5. Immediately try to pick up a paper clip with just the nail. What happens?

6. Repeat Steps 4 and 5 if you were unsuccessful.

7. Bang the nail on a hard surface before letting someone else try this test. (Be careful of the sharp point on the nail!) Record your observations in your Science Log.

What Happened?

Is the nail a magnet by itself? How about when it's next to your magnet? How about after you rubbed it with the magnet? What kind of magnet is the nail, a permanent or temporary magnet? How do you know?

Experiment 2: What Can Be Magnetized?

Try This!

Work in small groups to complete the experiment below.

MATERIALS NEEDED

- *same materials used in the previous activity*
- *additional objects, some nonmetal and some made from different metals*

Procedure:

1. Follow the same procedure that you used in the previous experiment. Try to magnetize other metal and non-metal objects. Record your findings in your Science Log.

What Happened?

Were you able to create magnets from other items? What does that say about the atoms in those items?

MAGNETIZE/DEMAGNETIZE

Name _____

Science Log

Use this log sheet to record the findings from your nail and magnet experiments.

Magnetizing a Nail ···

Question: Why are paper clips attracted to the magnet? _____

_____

What I did: _____

_____

What happened: _____

_____

Why it happened: _____

_____

What can be Magnetized? ···

Record your findings on the following chart.

Object Tested	Can it be Magnetized? (Yes or No)

Lesson 8: The Earth as a Magnet

Use this page when you introduce the concept of the Earth's magnetism to your students. The fun facts can be used to draw your students into the topic.

They'll Need to Know ...

The Earth has a thick iron and nickel core, which is constantly moving. This movement has caused our planet to develop a magnetosphere that extends into space, thus making our Earth similar to a huge magnet. The center of this magnet, the core of the Earth, allows travelers to use a compass for navigation and directional purposes. The magnet that surrounds the Earth does not point exactly North, but is slightly off course. The North Pole or "true North" and the Magnetic North are offset slightly because of the Earth's tilt. The compass needle will always point toward the Earth's "magnetic North."

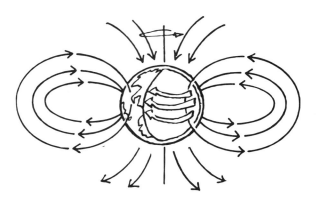

Prove It!

Discuss the idea of the Earth being a huge magnet and use a compass to show this. Use a magnet to move the compass needle around. Remove the magnet and the needle returns to its original position. Be sure students understand the difference between true North and magnetic North. Use a compass to find the four directions in your classroom. Label them so students can use them for their experiments.

Experiment 1: Test the Earth's Magnetism Teaching Notes:

In this experiment, the magnet should have come to rest, aligning itself with Earth's magnetic poles. Encourage students to use a compass or classroom direction signs to check. If there are not enough places to hang the magnet in this experiment, you could use a meter stick on top of two chair backs. Just make sure the chairs are pulled out as far as possible.

Experiment 2: Make a Compass Teaching Notes:

In this experiment, students temporarily magnetized the needle, which helped in creating a homemade compass that reacted to the Earth's magnetic field.

Journal Prompt

Now that you know that a compass always points to magnetic North, how do you think people use compasses to find their way around? Write about your ideas.

Homework Idea

Have students try testing other materials at home to make their own compass. Ask them to record the ones that worked best and write about their results or make a drawing of it.

FUN FACTS!

- The magnetic field of the Earth has switched several times in the past hundreds of thousands of years.

- The North Pole of the Earth is really the South Pole of the "Big Earth Magnet" because the north side of the compass needle is attracted to its opposite pole, which is south.

Experiment 1: Test the Earth's Magnetism

Try This!

Work in pairs to complete the following experiment. Record your findings on your **Test the Earth's Magnetism** Science Log.

Procedure:

1. Tie a piece of thread to the center of the bar magnet. With your teacher's help, suspend it from a ceiling light or other high object. Avoid metal objects.

2. When the magnet has come to a complete stop, note the direction it is facing.

3. Give the magnet a slight spin and wait for it to stop again. Note the direction it is facing. Record your observations in your Science Log.

• **bar magnet**
• **thread**
• **access to ceiling fixture or other object to attach your thread**

What Happened?

Where did the magnet point to when it stopped? Use a compass to check.

Experiment 2: Make a Compass

Try This!

Work in pairs to complete the experiment below. Record your findings on your **Make a Compass** Science Log.

Procedure:

1. Run a magnet over the needle 20 to 40 times, always stroking it in the same direction.

2. Place the needle carefully on top of the cork. Be careful not to stick yourself!

3. Float the cork and needle in your bowl of water.

4. Allow the water's surface to become still. What happens to the homemade compass?

• **sewing needle, about 1 inch long**
• **small bar magnet**
• **small piece of cork about 1/4 inch thick**
• **small glass bowl of water**

5. Turn the needle in a different direction, and then release it. What happens?

6. If you want to experiment further, try placing a magnet near your compass and watch what happens. How close does the magnet have to be to cause an affect? Record your observations in your Science Log.

What Happened?

In what direction did the needle point? Use a compass to check. What happened when you placed a magnet near your homemade compass?

Name _____

Science Log

Use this log sheet to record the findings from your compass experiments.

Test the Earth's Magnetism

Question: What are some things that might affect the direction of the hanging magnet? Why?

_____

_____

What I did: _____

_____

What happened: _____

_____

Why it happened: _____

_____

Make a Compass

Question: What are some other materials you could use to make a compass?

_____

What I did: _____

_____

What happened: _____

_____

Why it happened: _____

_____

Lesson 9: Electromagnets

Use this page when you introduce Electromagnets to your students. The fun facts can be used to draw your students into the topic.

They'll Need to Know ...

Any moving electric charge creates a magnetic field around it. A loop of wire with a current creates a magnetic field through the loop. You can increase the strength of this field by piling up a lot of loops. The more loops, the stronger the magnet. This is an electromagnet.

A needle sitting by a coil carrying current will also start to have a magnetic field. Ordinarily, all of the atoms in the needle point in all different directions, so the needle has no overall magnetism. But if you bring a needle near the south pole of your electromagnet, the north poles of the iron atoms will be attracted to the south pole of the electromagnet. As a result, they will all line up pointing in the same direction. The needle is now magnetized, with the north poles of its atoms facing the south pole of the electromagnet. The opposite poles attract each other, and the needle is drawn into the electromagnet. When the direction of current is reversed, the poles of the electromagnet reverse.

The principle of magnetic attraction is used to make a variety of devices, from doorbells (in which an iron rod is drawn into a coil to strike a chime) to pinball machines (in which current goes through a coil, drawing in a rod that is attached to the flipper), to the starter switch on a car.

By wrapping coils of wire around a piece of iron, such as a nail, one can create an electromagnet, which can be much stronger than a permanent magnet and have the ability of being turned off and on when needed. This is particularly helpful in the scrap iron industry, where an electromagnet is used to pick up and move large piles of iron.

Prove It!

Divide students into groups depending upon the availability of materials. These experiments will prepare students for the experiments in Lesson 10. It is important that they understand the idea of how a magnet is formed from the coils of wire and current running through them. Be careful! The wires can get very hot because they act as a short circuit. You don't want to keep them attached to the battery for too long. Review the directions for the experiments first to clear up any difficulties.

Experiment 1: Magnetic Attraction Teaching Notes:
In this experiment, use the largest nail possible that will fit inside the tube. Also, be sure to use a 6-volt battery. An ordinary 1.5-volt D-size battery will work, but it may go dead very quickly and will require more coils to get the same affect. Students will notice in this experiment that a coil of wire with a current flowing through it creates a magnetic field. Although the coil was not touching the iron nail, the coil's magnetic field attracted the nail with enough force to draw it into the tube. By reversing the connection, a magnetic field with opposite poles was produced, and this attracted the iron nail again.

Experiment 2: Make a Basic Electromagnet Teaching Notes:
In this experiment, the current flowing through the wire created a magnetic field around the nail. This turned the nail into a magnet. Adding coils produced a stronger magnetic force.

Experiment 3: Go Further with Electromagnets Teaching Notes:
In this experiment, students will notice that a larger nail and longer wire do not affect the results. However, winding the wire additional times should make the magnet stronger.

Journal Prompt

Magnetic strips are buried beneath roads at streetlights. How might they be important to the function of the streetlights?

Homework Idea

Have students research one device that uses electromagnets to function. Encourage them to make a display that shows what they've discovered.

- Joseph Henry built the first electromagnet in 1829.

- Having a "magnetic personality" means that people are naturally drawn to you.

Experiment 1: Magnetic Attraction

 **MATERIALS NEEDED**

- **insulated bell wire**
- **6-volt D-size battery**
- **wire stripper**
- **large iron nail, sharp end dulled with a file**
- **plastic or cardboard tube 4 to 6 inches (10 to 15 cm) long, about 1/4 inch (.6 cm) in diameter**
- **alligator clips**

Try It!

Work in small groups to complete the following experiment. Record your findings on your **Magnetic Attraction** Science Log.

Procedure:

1. Tightly wrap as many coils of wire as possible around the tube. Leave the two ends free.

2. Use a wire stripper to strip about 2 cm of insulation off the ends of the wire.

3. Insert the nail partway into the tube. Briefly connect the ends of the wires to the battery. (Leaving the wires connected too long can result in death for your battery and a burn for you from the hot wires.) What happens?

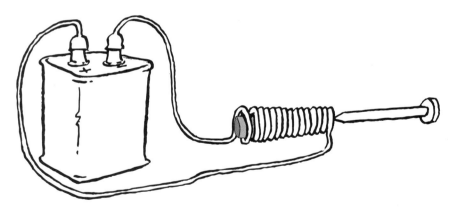

4. Predict what will happen if you reverse the connections to the battery and repeat the experiment. Try it. Record what happens.

What Happened?

What happened to the nail in this experiment? What caused it to react that way? What was the wire's role in this experiment?

Experiment 2: Make a Basic Electromagnet

MATERIALS NEEDED

- *16D iron nail*
- *piece of insulated copper wire about 8 inches (20 cm long)*
- *wire strippers*
- *battery holder with D-size batteries*
- *alligator clips*
- *paper clips*

Try This!

Work in pairs to complete the following the experiment. Record your findings on your **Make a Basic Electromagnet** Science Log.

Procedure:

1. Use wire strippers to strip about 2 cm of insulation off the end of the wire.
2. Tightly wrap the copper wire around the nail, making about five or six turns. The coiled wire should cover about two-thirds of the nail. Keep the ends of the wire available to connect to the alligator clips.
3. Connect the alligator clips to each stripped end of the copper wire.
4. Connect the other ends of the alligator wire to the battery holder's leads.
5. Try to pick up paper clips with the pointed end of the nail. What happens?
6. Try winding the copper wire more times around the nail, then do the test again. What happens?

What Happened?

What happened to the nail in this experiment? Did it pick up the paper clips? What happened when you added more coils to the nail?

Experiment 3: Go Further with Electromagnets

Try This!

Work in pairs to complete the experiment below. Record your findings on your **Go Further with Electromagnets** Science Log.

Procedure:

1. Follow the same procedure as you did for the previous activity. This time, try these variations: 1) different-sized nails, 2) different lengths of wire, and 3) different number of winds around the nail. Change one variable at a time.

MATERIALS NEEDED

- *nails, various sizes and materials*
- *insulated copper wire, various lengths*
- *wire strippers*
- *battery holder and 2 D-size batteries*
- *paper clips*

What Happened?

Does a larger nail and longer wire affect your results? What about winding the wire additional times around your nail? Why do you think that is?

64

ELECTROMAGNETS

Name _____

Science Log

Use this log sheet to record the findings from your experiments on electromagnets.

Magnetic Attraction ··

Question: What causes the nail to get sucked into the tube?

_____

What I did: _____

_____

What happened: _____

_____

Why it happened: _____

_____

Make a Basic Electromagnet ···································

Question: What happens when you wind more coils around the nail?

_____

What I did: _____

_____

What happened: _____

_____

Why it happened: _____

_____

_____

Name _____

Science Log

Use this log sheet to record the findings from the **Go Further with Electromagnets** experiment.

Question: If the electric current were to suddenly stop flowing through the electromagnet, what do you think would happen?

_____

_____

_____

Record the results of your experiment in the table below.

Size of Nail	Length of Wire	# of Winds	# of Paper Clips Picked Up
1.			
2.			
3.			

Explain how an electromagnet works:

_____

_____

_____

Lesson 10: Electric Motors

Use these pages when you introduce Electric Motors to your students. The fun facts can be used to draw your students into the topic.

They'll Need to Know ...

The magnetic field of a magnet exerts a force on an electric current flowing in a wire. A wire will move up or down, depending on the direction of the current and the direction of the magnet's magnetic field. The force that a magnet exerts on a current-carrying wire, called the deflecting force, is what drives most electric motors.

In the electric motor shown here, the permanent magnets exert forces on the electrical currents flowing through the loop of wire. When the loop of wire is in a vertical position, the forces on the top and bottom wires of the loop will be in opposite directions. These oppositely directed forces produce a twisting force, or torque, on the loop of wire that will make it turn. (Note: Experiment 1 is based on this model.)

Prove it!

These two experiments are cumulative. Students will have some experience from the previous lesson on electromagnets to complete these. Be sure to try the experiments yourself first, so you can help students with their questions. As for equipment, you can find disk magnets, ceramic magnets, and flexible wire at an electronic supply store. If supplies are limited, you might consider doing the second experiment as a class. Also, make photocopies of the diagram on page 68 for students to use as reference when doing **Experiment 2.**

Be careful! The wires can get very hot in these experiments because they act as a short circuits. Follow the instructions carefully. Encourage students to make minor modifications if their motors don't seem to work quite right.

Experiment 1: Move the Wire Teaching Notes:
In this experiment, the current flowing through the wire is affected by the magnetic field of the disk magnets. The wire will move up or down, depending on the direction of the current and the direction of the disks' magnetic field.

Experiment 2: Build an Electric Motor Teaching Notes:
In this experiment, student will see how current flows through a wire coil to create an electromagnet. Discuss with students how one face of the coil becomes a north pole and the other becomes a south pole when it's magnetized. The permanent magnet then will become attracted to its opposite pole on the coil and repelled to its like pole, causing the coil to spin. The blackened side of the wire insulates the wire for half a turn. As the coil spins and reaches the insulated portion of the wire, the current is turned off for a split second. This gives the coil a chance to avoid getting "stuck" north-to-south with the disk magnet. If this happened, the motor would just stop after one turn.

Journal Prompt

If all cars used electric motors instead of gasoline engines, what affect would that have on us?

Homework Idea

Ask students to find out what a commutator is, what it does, what the armature of an electric motor is used for, and what parts of their experiment correlate with these two parts.

FUN FACTS!

• There are magnets in electric motors that run on electricity. Electricity can also be made by turning the magnet in the motor.

• The first electric motors were constructed in 1821 by Michael Faraday in England and improved in 1831 by Joseph Henry in the United States.

ELECTRIC MOTORS

Diagram of an Electric Motor

Use this page when you complete **Experiment 2: Build an Electric Motor.**

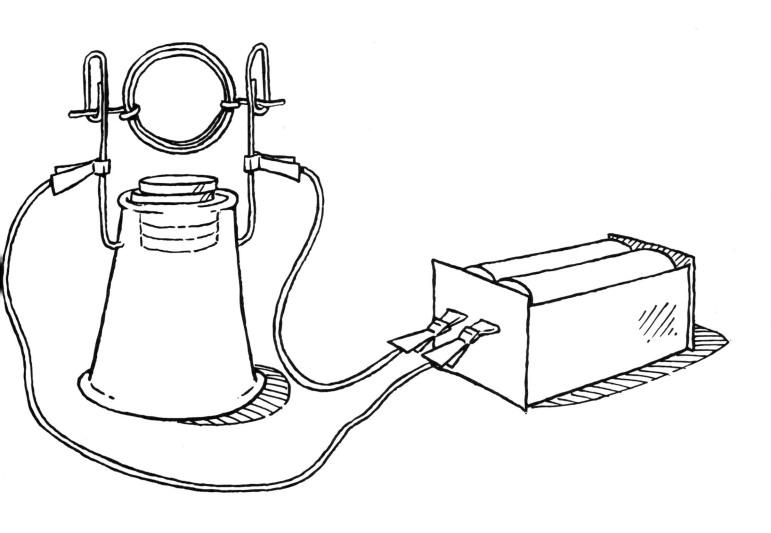

ELECTRIC MOTORS

Experiment 1: Move the Wire

MATERIALS NEEDED

- *ruler*
- *4 to 6 small disk magnets*
- *1 or 2 1.5-volt flashlight batteries*
- *wire strippers or sandpaper*
- *masking tape*
- *wood board approximately 2" x 4" x 6" (5 x 10 x 15 cm)*
- *approximately 2 feet (.6 cm) of flexible wire*

Try This!

Work in small groups to complete the following experiment. Record your findings on your **Move the Wire** Science Log.

Procedure:

1. Stack two or three magnets together so they stick together. Turn the stack on its side and tape it down on the board, just off-center.

2. Stack the remaining magnets together the same way. Turn the stack on its side and tape it down on the board, across from the first stack. Leave a gap of about 1/2 inch (1.3 cm) between the faces of the two magnet groups. A north pole should face a south pole.

3. Place the battery away from the magnets.

4. Remove the insulation from the ends of the wire. Loop the wire.

5. Feed one large loop of wire through the gap between the magnets. Be sure the ends of the wire will reach the battery, so you can connect them.

6. Watch the wire as you complete the circuit by touching the wire ends to the battery. What happens? What happens when you break the circuit? Be sure not to keep the circuit completed for long, as it can drain your battery very quickly and burn your fingers.

What Happened?

What happened to the wire? Why do you think that occurred?

Experiment 2: Build an Electric Motor

MATERIALS NEEDED

- **copy of motor diagram on pg. 68**
- **5 small disk or rectangular ceramic magnets**
- **2 large paper clips**
- **plastic or paper cup**
- **solid insulated 20-gauge copper wire, about 2 feet (60 cm) long**

- **masking tape**
- **6-volt lantern battery or regular flashlight batteries with a battery holder**
- **2 electrical lead wires with alligator clips at both ends**
- **wire strippers**
- **black permanent marker**

Try This!

Work in groups to build an electric motor that spins a wire coil. Use the diagram your teacher gives you for help with each step. Record your findings on your **Build an Electric Motor** Science Log.

Procedure:

1. Wind the copper wire into a coil about 1 inch in diameter. Make four or five loops.

2. Wrap the ends of the wire around the coil a couple of times on opposite sides to hold the coil together.

3. Leave 2 inches (5.1 cm) of wire from each side of the coil, and cut off any extra.

4. Strip the insulation off the ends of the wire from from the coil.

5. Blacken the top half of one side of the stripped wires with the marker.

6. Turn the cup upside down and place two magnets on top in the center.

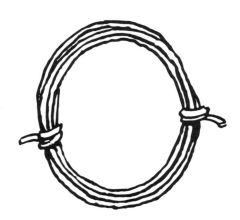

Experiment 2: Build an Electric Motor (continued)

7. Attach three more magnets inside the cup, directly beneath the original two magnets. This will create a stronger magnetic field as well as hold the top magnets in place.

8. Unfold one end of each paper clip. Tape the straightened ends of each paper clip to opposite sides of the cup.

9. Rest the stripped ends of the coil in the cradles formed by the paper clips.

10. Adjust the height of the paper clips so that the coil hangs about l/l6 inch above the magnets. It should be able to spin freely.

11. Adjust the coil and the clips until the coil stays balanced and centered while spinning freely on the clips. Good balance is important in getting the motor to operate well.

12. Trim off any excess wire that sticks out beyond the paper clip cradles.

13. Place a battery in the battery holder if using D-size batteries.

14. Use the alligator clips to connect one wire to each paper clip and to one terminal of the battery (or lead on the battery holder).

15. Give the coil a spin to start it turning. If it doesn't keep spinning on its own, check to make sure that the coil assembly is well balanced when spinning.

16. Check that the coil and the magnets are close to each other but do not hit each other.

17. Keep making adjustments until the motor works. Have patience!

What Happened?

What happened in this experiment? Why did the coil spin? What happened when the coil spun and reached the insulated portion of the wire? Why do you think that occurred?

Name _____

Science Log

Use this log sheet to record findings from the electric equipment experiments.

Move the Wire ··

Question: What causes the wire to move up or down? _____

_____

What I did: _____

_____

Why it happened: _____

_____

Build an Electric Motor ···

Question: What causes the coil to spin? _____

_____

What I did: _____

_____

What happened: _____

_____

Why it happened: _____

_____

STEP ⑦
CROSS THE CURRICULUM

Math and Magnets & Electricity

There's no better way to enhance learning and make it relevant to students than to tie it with all areas of the curriculum. In this step, you'll find a few fun curriculum-extending activities you might want to try! These pages are meant for you to photocopy and distribute to your students.

This page is filled with ways you can extend the learning to **Math**.

1. Measure with a Voltmeter

Using a voltmeter is an easy way to incorporate math into your electricity lessons.

1. Make a simple circuit with a light bulb.
2. Test the voltage across the bulb with the voltmeter.
3. Add another bulb and test the voltage.
4. Keep adding bulbs and testing the voltage. Predict what will happen before each test.
5. Record all results in a table.
6. Make a graph showing how the voltage changes as more bulbs were added.

2. Energy Consumption at Home

Graph energy consumption of households, and compare the ratio of electrical appliances, lights, and square footage to consumption. Have your parents help you determine the total square footage of your home. Look at a recent electric bill and record the total kilowatt hours used. Create a ratio of kilowatt-hours to total square footage. Graph the results. Could a large home have a large amount of square footage and low consumption? How? What about a small home having large electricity consumption?

3. Electromagnet Strength

Repeat Lesson 9's **Go Further with Electromagnets**, but keep the nail size and voltage the same.

1. Start by winding the wire just three times around the nail.
2. How many paper clips can you pick up using just the point of the nail?
3. Wind the wire three more times around the nail. Repeat Step 2.
4. Keep adding wire around the nail (wind it three times each time) and repeating Step 2. Record your results in a table.
5. Make a graph that shows how much wire is wound around the nail to the number of paper clips the nail was able to pick up.

You can also try changing just the number of batteries used and compare the number of paper clips the electromagnet can pick up with each test. Graph the results.

Social Studies and Magnets & Electricity

This page is filled with ways you can extend the learning to **Social Studies**. Photocopy this page, cut by activity, and distribute to students.

1. Timeline Tells All
Make a timeline that shows the history of electricity. Include the discovery of electricity, the invention of the first light bulb, the first power plant, the first power lines, the first solar cell, the first windmill to generate electricity, and other important developments in the field.

2. History of Magnets
Research the first magnets and who discovered them. Then, write a story about that person and present it to your classmates. Be sure to explain how times have changed since then for fun!

3. Magnets Around the World
Research how other countries use magnets. Are they similar to the way that we use magnets? How do they differ?

4. What's in Your Dirt?
Different parts of the country have different amounts of iron in the soil. See how much your region has with this fun activity. Take a cupful of soil from the area and pour it in a plastic bag with a zipper. Then, take a cupful of water and pour it into the bag. Zip the bag up, and make sure it's locked tight. Then, take your bar magnet and rub the bottom of the bag with it. Turn the bag over. Those black flecks are actually iron in the soil!

5. Compass Adventure
Use a compass to plan a trip around the America. Write down all the directions and places you will visit. Now try it with the world. Where would you want to visit?

Language Arts and Magnets & Electricity

This page is filled with ways you can extend the learning to **Language Arts.** Photocopy this page, cut by activity, and distribute to students.

1. Electricity and Magnet Music

Write a rap that uses vocabulary and concepts on electricity and/or magnetism. Then, perform it into an audiotape cassette. You may want to work with a partner in coming up with great background music!

2. Electricity and Magnet Poetry

Write an electricity or magnetism Haiku. A Haiku is a poem with five syllables for the first line, seven syllables for the second line, and five syllables for the last line. Here is an example:

> *Magnets*
>
> Repels to like poles
> Attracts when they are unlike
> They are magnetic!

3. Write About Safety

Write a safety manual for working with electricity. Be sure to include everything your classmates should know when working with circuits and batteries.

4. Create a Magnet Care Poster

It's very important to care for your magnets properly. Create a poster explaining the dos and don'ts when working with magnets. Include a tips section on appropriate places to store magnets.

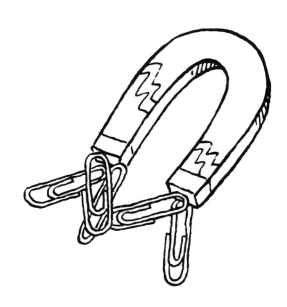

Reading, Art, and Magnets & Electricity

This page is filled with ways you can extend the learning to **Reading and Art.** Photocopy this page, cut by activity, and distribute to students.

1. Act as Author
Create an electricity and/or magnetism book with facts and pictures for younger readers. Use some of the experiments you did in this unit or find ones that are simpler. Read two children's books on electricity and magnetism to help you decide how to present information. Two good books are **Electricity and Magnetism** by Brian J. Knapp and **Electricity and Magnets** by Barbara Taylor.

2. Story Time
Find a great book about magnets and/or electricity, and read up on one of the topics you discussed in class. Then, show what you know by explaining the concept to your classmates.

3. My Favorite Inventions
Many inventors use circuits to make their inventions. Come up with a list of five inventions that you really like and read about the person who invented them on the web or at your local library. Were you surprised by anything you read? How would you change the inventions to fit your needs?

4. Lightning Art
On black construction paper, draw some interesting lightning bolts with white chalk.

5. More Magnet Mosaics
Make a whole class magnet mosaic on one of your metal-backed whiteboards, an extension of the bulletin board idea.

6. Make a Message Board
Use construction paper, laminating paper, and magnets to create a refrigerator message board. Use fancy designs for the border with lines or open space in the middle.

7. Electrifying Portraits
Work with a partner and take turns rubbing a balloon on your head. Draw what each of you looks like while your hair is positively charged.

8. Fun with Iron Filings!
Cut out a picture of a person from a magazine. (Make sure you have permission to cut it out before you do.) Place the picture in a shoebox top, and put iron filings on top of the picture. Use a bar or horseshoe magnet beneath the shoebox top to slide the iron filings on the person! Give him or her a mustache or funny hair.

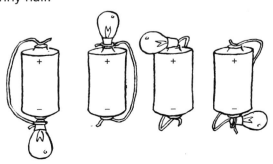

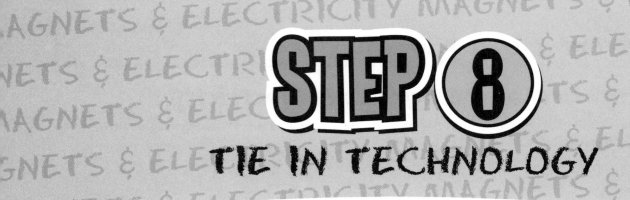

STEP 8
TIE IN TECHNOLOGY

Two Great Projects

Technology offers wonderful opportunities for reinforcing learning of all types. In this section you'll find two great projects that will allow you to take full advantage of all technology has to offer while at the same time strengthening the knowledge gained during the unit of study. Depending on the age group, these activities may be rather advanced. They can be simplified by not using technology or by working through the activities as a whole group. The options are limitless!

1. Create a Multimedia Presentation: Conserving Electricity

This unit of study concentrates not only on understanding electricity, but also on understanding our personal consumption of electricity with the electrical components we use at home. A multimedia presentation provides a great way for students to explain their understanding of personal electricity usage and ways to conserve electricity.

Discuss what you expect as far as content, such as how much electricity your students think they use each day, what the actual consumption is based on bills and resources, how to conserve electricity, and how resources should be cited in a bibliography.

Divide students into groups of two or three. Give groups time to brainstorm their portion of the presentation and then distribute the **Storyboard** worksheet on page 79. (The groups will more than likely need multiple copies.) If possible, allow the students to spend some time at the computer experimenting with design elements and searching for movies, photos, links or other elements they'd like to include in their presentation. Encourage the use of original artwork and sounds.

Distribute the **Multimedia Presentation Checklist** on page 78. Allow multiple work sessions for planning and the actual creation of the presentation. Then, plan a class "showing" of each group's presentation.

The computer tools you use will depend on what is made available to your school. Some programs that may enhance the project include many of the word processing and desktopping software on the market. Other tools include a digital camera and even an audiocassette tape. Another way to go is to create a poster per **Storyboard** and use the posters in the presentation. The choices are limitless. However, be sure the students are comfortable using the tools before they start. Also, when students present the project to the class allow them to use the computer to enhance the presentation.

MULTIMEDIA PRESENTATION CHECKLIST

Name _____

Planning

☐ Have I researched the topic and decided how to show it in a presentation?

☐ Have I located outside sources (graphics, sounds, links to web sites, and movies) to use within the presentation?

☐ Have I developed a **Storyboard**?

☐ Have I determined which tools I need to complete the task?

☐ Has each slide or card been designed and numbered?

Content

☐ Does my presentation clearly prove a point, explain something, or answer a question?

☐ Does the presentation support the content: not too silly if the subject is serious and vice versa?

☐ Did I include a table of contents or clear navigation?

☐ Are all my references properly cited on a bibliography or reference card?

☐ Did I include an "about the author(s)" card?

Design

☐ Is it easy to work through the presentation?

☐ Is there a good contrast between text color and background color?

☐ Are font choices consistent? (Try to use 3 font types or fewer.)

☐ Are the sounds, movies, and animations appropriate to the content?

☐ Is the text free of spelling, grammar, and punctuation errors?

☐ Are the graphics clear?

☐ Is the presentation interesting?

Presentation

☐ Have I rehearsed the presentation?

☐ Have I completed a "dry run" in front of others to make sure the presentation will run smoothly?

MULTIMEDIA PRESENTATION STORYBOARD

Name _____

Use these boxes as you're designing each screen for your presentation on conserving electricity.

1.

2.

3.

4.

5.

6.

This second project will allow you to take full advantage of all technology has to offer while at the same time strengthen the knowledge gained during the unit of study. The next few pages explain what items to include on a web page, but they do not explain how to set up a web site. Learning Resources, Inc. offers a wonderful book to explain how to do this. It's called **LER 2282 Technology in the Classroom: Web Page Creation.**

2. Create a Web Site: How an Electromagnet is Created

If your students have already experimented or are ready to learn about web page development, creating a web page is another great way to "show what they know." The steps in this book explain how to create a compelling web site on electromagnets. They do not provide directions on how to build the actual web site.

First, spend time viewing web sites. Discuss what makes an effective web site as well as what makes a poor web site. (Use the checklist on page 81 as a guide here.) Introduce the topic for your students' web development project, and divide them into groups of two or three.

Next, discuss what you expect in terms of content. Suggest covering the following points, but allow students the opportunity to provide their own input:

- Should they tell what an electromagnet is?
- Should they include each step in the creation of an electromagnet?
- Do you expect them to include other information such as how electromagnets are used in society, such as metal detectors and scrap yard electromagnet cranes?
- Should they include bibliography information?

Give students time to brainstorm their web site and then distribute the **Web Site Flow Chart** worksheet on page 82. If possible, allow the students to spend some time at the computer experimenting with design elements and searching for movies, photos, links or other elements they'd like to include as part of their web site. Encourage the use of original artwork and sounds.

Distribute the **Web Design Checklist** on page 81. Allow multiple work sessions for planning and for the actual creation of the web pages. If possible, post the sites to the school server to allow other classes within the school to view the pages. Give students ample time to view each group's site.

WEB DESIGN CHECKLIST

Name _____

- ☐ Is my site's objective clear?

- ☐ Is the subject divided with different subject matter on different pages?

- ☐ Is the text easy to read?

- ☐ Do all links work correctly?

- ☐ Have spelling and punctuation been checked on each page?

- ☐ Is navigation simple to use?

- ☐ Are there links at the bottom of each page so the user can navigate back to the top of the page, the home page, the table of contents, or related information on the subject?

- ☐ Is there a balance between graphics and text?

- ☐ Are font and point size consistent?

- ☐ Is the design consistent?

- ☐ Do all links work correctly?

WEB SITE FLOW CHART

Name _____

Use this flow chart to help you think through the design and structure of your web site. Provide notes on buttons, links, design elements, and content.

Page 1

Page 2

Home Page-Title Page

Page 3

Page 5

Page 4

STEP 9

ASSESS LEARNING

Introduction on Assessment

You've done your job. The content was incredible, the "hands-on" learning opportunities were abundant, and the delivery was no doubt sublime! Now let's see how much actual "learning" took place. There are a number of great ways to assess student learning. We've included some of these methods within the next few pages, complete with rubrics and actual assessments you can photocopy and have students take.

Tests

A well-written test is the granddaddy of all assessment tools. If you've included everything you want the students to know, a test can be a very reliable measure. We've included two types of tests for this unit: 1.) a Q&A test, and 2.) a multiple choice, matching, and True or False test.

Rubrics

Rubrics allow students and teachers to record their perceptions and opinions. Whenever using rubrics, it's important to encourage honest reporting on the students' part. We've included two rubrics in this section — one for the student and one for the teacher.

Journals

Requiring students to keep a journal as you study a topic serves two purposes:
1. It causes the student to recall the information they've just studied.
2. It helps you determine just how much information they took away from the lesson, which you can use to determine the concepts that need further discussion.

The sample journal page included in this book outlines the following areas:

1. *What we studied today.* This encourages students to recap the day's learning.
2. *My experience with this topic.* Students use this space to share their own experiences with the topic, such as their familiarity with electricity and magnets or the fact they have done the same activity before in another class. If students discuss the latter in this section, encourage them to write about what the activity demonstrates.

3. *Questions I still have.* This is an excellent area for you to identify what students do not understand or to take the learning to the next level. This space allows students to ask any questions they still have surrounding the subject.

Science Logs

Reading a student's **Science Log** will give you clear feedback on whether they understand the scientific concept associated with the experiment. Throughout the lessons in **Step 6**, we've included Science Logs for students to fill out when they conduct an experiment. Even though you might provide students with directions for completing each experiment, it's important for them to write down exactly what they did, what materials they used, what the results were, and what they feel the reasons were for the outcome. If what they write is correct and scientifically true, great! If not, you'll know what to review in your upcoming lessons.

A Note About Assessing Projects

While the projects in **Step 5: Plan a Project** provide a great way to reinforce learning, they can be tricky to assess — especially if the projects are group activities. In the case of a group project, always monitor each group's performance. Make sure each person is doing a fair amount of the work. If possible, include a peer assessment as part of the overall grade. It should be noted that projects don't always cover a complete topic, but rather portions of a topic. Therefore, never base a student's grade for the unit of study solely on a project. We have included some sample project assessment pages throughout this chapter on pages 85-86 for **Step Five: Plan a Project.**

MY ELECTRICITY/MAGNETISM JOURNAL

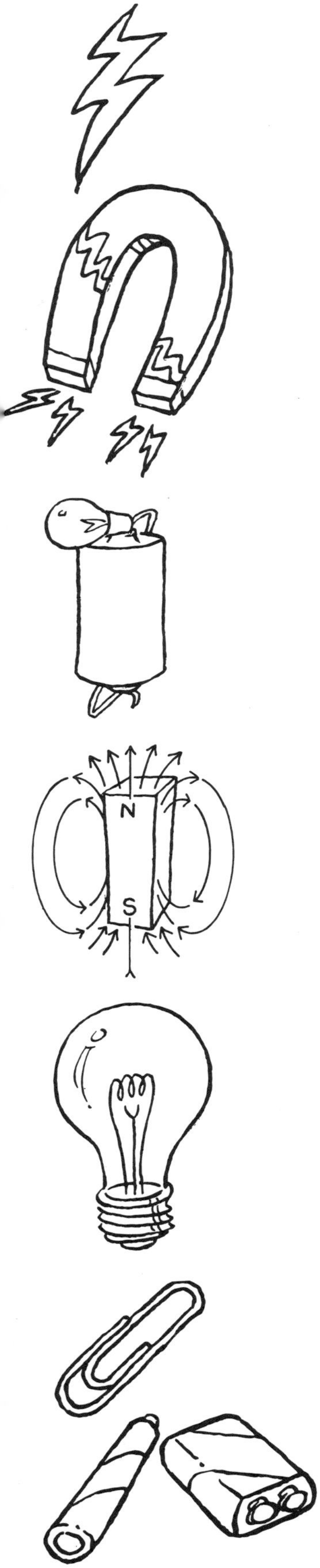

Name _____   Date _____

What we studied today:

_____

_____

_____

_____

_____

_____

_____

My experiences with this topic:

_____

_____

_____

_____

_____

_____

_____

Questions I still have:

_____

_____

_____

_____

_____

_____

_____

_____

PEER ASSESSMENT RUBRIC

Student-to-Student Assessment

Expectations	Actual Performance				
	Never	Sometimes	Frequently	Always	Points
My teammate was helpful.	1 point	2 points	3 points	4 points	
My teammate listened to the ideas presented and participated in group decisions.	1 point	2 points	3 points	4 points	
My teammate contributed a fair amount of work toward the final outcome.	1 point	2 points	3 points	4 points	
My teammate accepted criticism and redirection in a positive manner.	1 point	2 points	3 points	4 points	
Other	1 point	2 points	3 points	4 points	
				Total Points	

Evaluator's Name: _____

Comments: _____

_____

_____

Subject's Name: _____

Comments: _____

_____

_____

Teacher's Comments: _____

_____

PRESENTATION/PROJECT RUBRIC

Teacher Assessment

Expectations	Actual Performance & Point Assignment				
	Poor	Okay	Good	Great	Points
Organization	1 point	2 points	3 points	4 points	
Content	1 point	2 points	3 points	4 points	
Mechanics	1 point	2 points	3 points	4 points	
Design	1 point	2 points	3 points	4 points	
Presentation	1 point	2 points	3 points	4 points	
Other	1 point	2 points	3 points	4 points	
				Total Points	

Group Members: _____

Subject's Name: _____

Teacher's Comments: _____

_____

Organization: _____

_____

Content: _____

_____

Mechanics: _____

_____

Design: _____

_____

Presentation: _____

_____

Understanding Electricity and Magnetism Assessment Test

Name_____ Date _____

True or False

Read each sentence below. Write a T on the line if it is true or an F on the line if it is false.

1. The nucleus of an atom is negatively charged.   _____

2. Magnets have two poles, the East Pole and the West Pole.   _____

3. Wind energy is created through wind socks.   _____

4. Electrons flow from negative to positive in a battery.   _____

5. Any metal can be magnetized.   _____

Fill in the Blank

Fill in the blank to complete each sentence.

6. A(n) _____ has a negative charge.

7. A(n) _____ acts like a pump for electrons.

8. A(n) _____ is the unit that measures electrons moving through a circuit.

9. A(n) _____ easily allows electrons to move through it.

10. A(n) _____ slows the flow of electrons.

11. A(n) _____ can help you tell direction.

Multiple Choice

Circle the correct answer that finishes the sentence.

12. Protons are:
      A. neutral   B. positive   C .negative

13. An insulator:
      A. conducts   B. helps   C. blocks the flow of electrons.

14. The strong force holds together:
      A. electrons   B. protons   C. magnets

15. Static electricity is the build up of:
      A. protons   B. neutrons   C. electrons

16. Direct current means that the electrons flow:
      A. in both directions   B. in one direction   C. in neither direction

17. A circuit in which the electron flow moves directly from one pole of the battery to the other is called a:
      A. series circuit   B. short circuit   C. parallel circuit

TEACH TEN TERRIFIC LESSONS

Understanding Electricity and Magnetism Q & A Assessment

Name_____ Date _____

1. How does electricity affect magnetism?_____

_____

_____

2. What are three ways we can generate electricity?

   a.) _____

   b.) _____

   c.) _____

3. Explain how a piece of iron can become magnetized. _____

_____

_____

4. Explain how a compass works. _____

_____

_____

5. Explain what the strong forces does. _____

_____

_____

6. What is a potential problem with wind energy? With nuclear energy? _____

_____

_____

7. Explain what a magnetic field is. _____

_____

_____

Use the materials your teacher provides to build a simple, series, and parallel circuit. Be sure the light is activated in each circuit. Show your teacher your completed circuits.

Electricity and Magnetism Science Fair

It's been an interesting few weeks. You've worked hard to ensure student learning. You've required a lot of your students. Everyone, including you, knows a lot more today than you did a few weeks ago. It's time to celebrate your success! What better way to wrap up the unit than with a science fair?

1. **You'll Need to Get Organized**
   Look at all the different lessons, activities, and experiments you did for this unit and list them.

2. **You'll Need to Create Committees**
   Make sign-up sheets for each lesson, activity, and experiment; write the number of students it will take to make and run an experiment on that topic. Try to have enough slots on the sign-up sheets to include all students in your class. If there are more slots than students, combine some of the topics as you see fit.

   Explain to your students that you plan to have a science fair in which their job will be to showcase and teach what they have just learned to other students, teachers, the administration, community members, and parents. Tell students they will need to sign up for a topic they feel comfortable presenting and for which they are willing to create and run an accompanying display. Remind students there are only so many slots on each topic and that they may not get their first choice.

3. **You'll Need to Plan Presentations and Displays**
   After students have signed up for their topic, provide them with copies of the checklist on page 90 to help them with their planning. Provide them with materials needed to make props and display boards.

Help students get started if necessary, and keep them motivated throughout the planning process. Remind them they are doing this so they can show others what they have learned. Set a deadline for each stage of planning to keep everyone on task. It should not take any longer than one week to get the displays ready. Be sure to book the area where your displays are to be held.

4. **You'll Need to Advertise the Event**
   You will want to send out the invitations as soon as possible to give others time to plan for the event. Use the sample invitation printed on page 91. Also, have student volunteers make posters to place around school. Contact your local newspaper to advertise the time and date of your big event.

5. **You'll Want to Have one Last Practice Run**
   Have students show you that they can easily perform the experiment or do the demonstration. Then, set up the displays. Be sure to include materials generated by the activities in the unit. You are ready to go!

CELEBRATE!

Student Science Fair Planning List

Name(s) of student(s) planning display: _____

_____

Check off items on this list as you complete them.

☐ Sign up for a topic.

☐ Review the work you did in class with that topic.

☐ Write down all the information you learned.

☐ List any materials you will need.

☐ Use web sites or books to research your topic in more detail. Take notes.

☐ Write up a few blurbs to create posters such as "Did You Know ...," "Interesting Facts about ...," "What I learned ...," or "What is Happening..." and include them on your display board. Also include pictures of your group working with the equipment.

☐ Gather the materials you will need to demonstrate the concept for a group of people.

☐ Get a display board to use as a backdrop.

☐ Design your display board on white paper first. Choose a drawing from the group or combine ideas.

☐ Transfer the drawing onto the display board along with the posters and any pictures.

☐ Have the display checked by the teacher.

☐ You are ready!

You're Invited

A Science Fair!

Our class has just concluded an incredible unit on electricity and magnetism, and now we want to show you what we know!

What? A Science Fair on Electricity & Magnetism

When?

Where?

What Time?

ANSWER KEY

Page 11: Magnets & Electricity Vocabulary Practice

1. permanent magnet
2. conductor
3. resistor
4. ampere
5. electron

6. electricity
7. battery
8. circuit
9. compass
10. magnet

11. static electricity
12. nucleus
13. strong force
14. short circuit
15. protons

Page 12: Human Body Vocabulary Crossword Puzzle

Across
1. electron
2. resistor
3. circuit
4. current
5. magnetism

Down
6. resistance
7. ampere
8. electricity
9. battery
10. conductor
11. magnet
12. compass

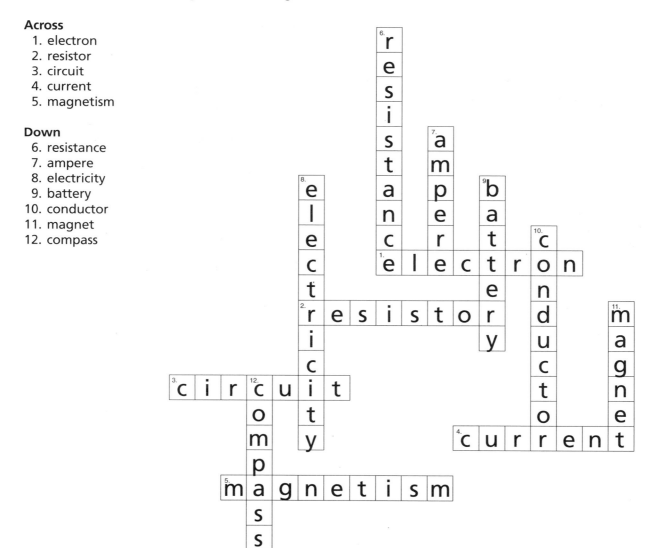

Page 55: Show What You Know: Magnets

1. poles
2. opposite
3. like
4. iron, nickel, or cobalt
5. magnetism 6. magnetic resonance imaging
7. float
8. answers may vary

Page 87: Understanding Electricity and Magnetism Assessment Test

1. F
2. F
3. F
4. T
5. F
6. electron
7. battery
8. ampere
9. conductor
10. resistor
11. compass
12. B. positive
13. C. blocks
14. B. protons
15. C. electrons
16. B. in one direction
17. B. short circuit

Page 88: Understanding Electricity and Magnetism Q & A Assessment

1. When electricity flows through a wire, a magnetic field is created. When this same wire is wrapped around an iron object, an electromagnet is produced.

2. Examples: solar, wind, hydroelectric, geothermal, oil, coal, natural gas, ocean and nuclear.

3. The atoms in the iron line up when brought close to a magnetic field. The lining up of these atoms creates a magnetic field and a magnet.

4. The compass needle is actually a magnet, as is the Earth. The north pole of the compass is the needle and it points to the magnetic North, which is the South Pole of the Earth magnet.

5. Since the protons in a nucleus are positively charged and like charges repel, they need a strong force to keep them together. This is the strong force of the atom.

6. Answers will vary. Possible answers: There may be no wind. There may be a problem with what to do with the toxic waste.

7. It is the area around a magnet in which an object can be magnetized. The stronger the magnet, the larger the magnetic field.